T0224040

Parametric Continuation and Optimal Parametrization in Applied Mathematics and Mechanics

Parametric Continuation and Optimal Parametrization in Applied Mathematics and Mechanics

by

V.I. Shalashilin
Moscow Aviation Institute,
Moscow, Russia

and

E.B. Kuznetsov
Moscow Aviation Institute,
Moscow, Russia

SPRINGER-SCIENCE+BUSINESS MEDIA, B.V.

A C.I.P. Catalogue record for this book is available from the Library of Congress.

ISBN 978-90-481-6391-5   ISBN 978-94-017-2537-8 (eBook)
DOI 10.1007/978-94-017-2537-8

Printed on acid-free paper

Contents

Preface

A decade has passed since *Problems of Nonlinear Deformation*, the first book by E.I. Grigoliuk and V.I. Shalashilin was published. That work gave a systematic account of the parametric continuation method. Ever since, the understanding of this method has sufficiently broadened. Previously this method was considered as a way to construct solution sets of nonlinear problems with a parameter. Now it is clear that one parametric continuation algorithm can efficiently work for building up any parametric set. This fact significantly widens its potential applications.

A curve is the simplest example of such a set, and it can be used for solving various problems, including the Cauchy problem for ordinary differential equations (ODE), interpolation and approximation of curves, etc. Research in this area has led to exciting results. The most interesting of such is the understanding and proof of the fact that the length of the arc calculated along this solution curve is the optimal continuation parameter for this solution.

We will refer to the continuation solution with the optimal parameter as the best parametrization and in this book we have applied this method to variable classes of problems: in chapter 1 to non-linear problems with a parameter, in chapters 2 and 3 to initial value problems for ODE, in particular to stiff problems, in chapters 4 and 5 to differential-algebraic and functional differential equations.

The fact that the transition to the best parameter in the Cauchy problem for ODE of normal form can be made using an analytical transformation, which we will call λ-transformation, has surprised even us.

Another notable result, which we discuss in chapter 6, is the development of a general approach to applications of the best parametrization in parametric approximation problems.

In chapter 7 we will consider potential applications of parametric continuation to building more complex one-parametric sets, i.e. solution sets for non linear boundary problems for ODE with a parameter.

Finally, in chapter 8, we will show how to use the best parametrization for the continuation of a solution in a neighborhood of singularities.

The authors express their gratitude to N.S. Bakhvalov and G.M. Kabel'kov for their attentive and favorable discussions of these results. We specially thank V.A. Trenogin and V.V. Dikusar, who undertook the hard job of reading this monograph and made very useful notes. We express our special thanks to the untimely deceased V.V. Pospelov for his support at all stages of this work. We also thank D.V. Shalashilin and T. Yu. Shalashilin for their inestimable assistance when translating this book into English, and our colleague A.V. Deltsova and student S.D. Krasnikov for their assistance when preparing the camera-ready manuscript.

The principle scientific results covered in this book were obtained with the financial support of the Russian Foundation of Basic Research (Project Codes 01-01-00038 and 03-01-00071).

Chapter 1

NONLINEAR ALGEBRAIC OR TRANSCENDENTAL EQUATIONS WITH A PARAMETER

Solutions of many typical mathematical problems are often given by sets depending continuously on one parameter and differentiable with respect to this parameter. A continuous smooth curve in a multi-dimensional space is the simplest example of such a set. For instance, it can be a solution of a system of nonlinear equations with parameter. Other examples are integral curve of Cauchy problem for a system of ordinary differential equations (ODE), interpolating or approximating curve, etc.

In this chapter we will consider a process of construction of the curve as a solution of nonlinear equations system containing parameter. The approach will allow us to demonstrate the idea of parametric continuation method in a historically successive way, because the method itself was originally formulated just for such problems.

A parametric set, which is a solution of boundary-value problem for the ODE system with a parameter, serves as a more complicated example of continuous and differentiable set. It will be considered in the chapter 7.

1. Two forms of the method of continuation of the solution with respect to a parameter

We consider a system of n nonlinear algebraic or transcendental equations in n unknowns $x_1, x_2, \ldots, x_n$, containing a parameter p. In n - dimensional Euclidean space ^{n} the system can be represented in form

$$F(x, \ p) = 0. \tag{1.1}$$

1

Here $x = (x_1, x_2, \ldots, x_n)^T$ is a vector and $F = (F_1, F_2, \ldots, F_n)^T$ is a vector-function in space ^{n}.

We are interested in the behaviors of solution of the system (1.1) when the parameter p varies. Let the solution $x_{(0)} = (x_{1(0)}, x_{2(0)}, \ldots, x_{n(0)})$ of equation (1.1) to be known for a certain value of $p = p_0$, i.e.

$$F(x_{(0)}, \ p_0) = 0. \tag{1.2}$$

We introduce the space $^{n+1} : \{x, \ p\}$, augmenting the space ^{n} with the coordinate axis along which the parameter p is measured, and consider a neighborhood U of the point $(x_{(0)}, \ p_0) \in ^{n+1}$ in form of a rectangular parallelepiped centered at the point $(x_{(0)}, \ p_0)$. Properties of the solution of system (1.1) in this neighborhood are established by the well - know implicit function theorem (see., e.g., [40]). It proves that if the following conditions are satisfied:

1) vector - function F (i.e., all its components F_i, $i = \overline{1, \ n}$) is defined and continuous in U;

2) in U there exist partial derivatives of F_i $(i = \overline{1, \ n})$ with respect to all arguments x_i $(i = \overline{1, \ n})$ and the parameter p and they are continuous;

3) the equation (1.1) is satisfied in the point $(x_{(0)}, \ p_0)$, i.e., the equality (1.2) is fulfilled;

4) the Jacobian $\det(J)$ of the vector - function F is not zero at the point $(x_{(0)}, \ p_0)$ then the solutions x_i $(i = \overline{1, \ n})$ of system (1.1) in some neighborhood of the point $(x_{(0)}, \ p_0)$ are single - valued continuous functions of p

$$x_i = x_i(p), \qquad i = \overline{1, \ n} \tag{1.3}$$

Thus, $x_i(p_0) = x_{i(0)}$ $(i = \overline{1, \ n})$ and the derivatives dx_i/dp $(i = \overline{1, \ n})$ are also continuous in this neighborhood.

The Jacobian of the vector - function F is a determinant $\det(J)$ of its Jacobi matrix J

$$J = \frac{\partial F}{\partial x} = \frac{\partial(F_1, \ldots, F_n)}{\partial(x_1, \ldots, x_n)} = \left[\frac{\partial F_i}{\partial x_j}\right] = \begin{bmatrix} \dfrac{\partial F_1}{\partial x_1} & \dfrac{\partial F_1}{\partial x_2} & \cdots & \dfrac{\partial F_1}{\partial x_n} \\ \dfrac{\partial F_2}{\partial x_1} & \dfrac{\partial F_2}{\partial x_2} & \cdots & \dfrac{\partial F_2}{\partial x_n} \\ \vdots & \vdots & \ddots & \vdots \\ \dfrac{\partial F_n}{\partial x_1} & \dfrac{\partial F_n}{\partial x_2} & \cdots & \dfrac{\partial F_n}{\partial x_n} \end{bmatrix},$$

$$\tag{1.4}$$

$$i, j = \overline{1, \ n}.$$

Thus, the implicit function theorem establishes that the solution of system (1.1) in some neighborhood of the point $(x_{(0)}, p_0)$ forms a unique curve K which has a parametric representation (1.3) and passes through the point $(x_{(0)}, p_0)$ if conditions 1 – 4 are satisfied.

To obtain a solution $x_{(1)}$ of the system (1.1) for a value p_1 close to p_0, we can move along curve K. Of course the point $(x_{(1)}, p_1)$ must remain in the close neighborhood of $(x_{(0)}, p_0)$. In other words, we can uniquely continue the solution from the point $(x_{(0)}, p_0)$ within the neighborhood. If conditions 1 – 4 are fulfilled in some neighborhood of the point $(x_{(1)}, p_1)$ the solution can be continued again, and so on. Thus, the conditions 1 – 4 are sufficient for the solution of the system (1.1) to form a continuous curve K in space $^{n+1}$. This enables one to obtain solutions $(x_{(k)}, p_k)$ by moving along this curve from the known solution $(x_{(0)}, p_0)$ as illustrated on the Fig. 1.1 in a three - dimensional space ^{3}: $\{x_1, x_2, p\}$. This process just realizes the method of parametric continuation of a solution.

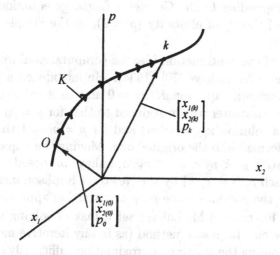

Figure 1.1.

The conditions 1 – 3 are not too restrictive and are fulfilled in most applied problems. Those points at which the condition 4 is also satisfied, i.e. $\det(J) \neq 0$, will be called regular. Those at which $\det(J) = 0$ will be called singular. As will be seen later, the parametric continuation can be performed not only at regular but also at singular points. However, bifurcation of the curve K of the set of solutions of the system may appear (1.1) at singular points.

The idea of solution continuation has long been known and exploited in mathematics and mechanics. Suffice is to say that this very idea actually underlies the well - known perturbation method (the small parameter method) whose first applications go back to works of U.J.J. Leverier (1856) and H. Poincare (1892).

The continuation has repeatedly been used to prove the existence of solutions of nonlinear equations. The idea is as follows. A parameter p is introduced into the primary equation so that for initial value of the parameter $p = p_0$ the solution of the equation is known, and the equation is converted into primary equation when $p = p_k$. After that the proof of existence of a solution is reduced to the proof of existence of continuous curve K. For more details see review [38]. In the theory of finite deflection of plates the above procedure had been successfully applied by N.F. Morozov [82, 83, 84, 85, 86]. He introduced a parameter p as a multiplier to nonlinear parts of operators in the Foppl - Karman equations and proved topological homotopy of the operators of the equations for $p = 0$ and $p = 1$. Thus the parameter p was used for construction of a continuous topological transformation from the linear operators, corresponding to the Germen - Lagrange equations and the plane problem of theory of elasticity ($p = 0$), to the Fopple - Karman equations ($p = 1$).

The first use of the continuation idea for computational purposes appears to be due to M. Lahaye [70] (1934). He introduced a parameter p into a transcendental equation $H(x) = 0$ and thus reduced it to the form (1.1). The parameter was introduced so that for $p = p_0 = 0$ it was easy to obtain a solution $x_{(0)} = x(p_0)$ and for $p = p_k = 1$ the equation would be transformed into the original one. Moving the sequence of parameter values $p_0 < p_1 < p_2 < \ldots < p_k$ M. Lahaye proposed to construct a solution for each p_i $(i = \overline{1, \ k})$ by the Newton - Raphson method using the solution for the previous value p_{i-1} as a starting approximation.

Formulating his process M. Lahaye set a task of solving the central problem in Newton - Raphson method (as in any iterative method), i.e. a problem of choosing the starting approximation sufficiently close to the desired solution. Indeed, if there are no singular points on the interval $p_0 \leq p \leq p_k$ on the curve K of the set of solutions of equation (1.1) it is always possible to choose the motion step for the parameter p so small that the desired solution $x(p_i)$ and its starting approximation $x(p_{i-1})$ are sufficiently close to each other to meet the convergence conditions for the Newton - Raphson method with respect to the choice of the starting approximation. This follows from the continuity and smoothness of the curve K. Later in his paper [71] M. Lahaye extended this approach to systems of equations.

Obviously M. Lahaye's suggestion is also applicable to equations which already contain a parameter. For these such an approach enables one to design a process, stepwise with respect to a parameter, for constructing a set of solutions in the parameter range of interest $p_0 \leq p \leq p_k$. Let $x_{(i)}^{(j)} = x^{(j)}(p_i)$ denote an approximate value of the desired solution $x_{(i)} = x(p_i)$. The process suggested by M. Lahaye for constructing the solution of (1.1) by going from p_{i-1} to p_i can then be written as

$$x_{(i)}^{(0)} = x_{(i-1)},$$

$$x_{(i)}^{(j)} = x_{(i)}^{(j-1)} - J^{-1}\left(x_{(i)}^{(j-1)}, p_i\right) F\left(x_{(i)}^{(j-1)}, p_i\right), \qquad (1.5)$$

$$j = 1, \, 2, \, \ldots,$$

as long as $\|x_{(i)}^{(j)} - x_{(i)}^{(j-1)}\| > \varepsilon$.

Here $\varepsilon > 0$ is pre - assigned error in the norm of the desired solution; $\|x\|$ is a norm of the vector x; $J\left(x_{(i)}^{(j-1)}, p_i\right)$ is the Jacobi matrix of vector function F for $x = x_{(i)}^{(j-1)}$ and $p = p_i$.

In our opinion the importance of M. Lahaye's works is that he gave an example of constructing a stepwise process with respect to a parameter which implements the main principle of the continuation method, i.e. to use at each step the information about the solution obtained at the previous step (steps). From this point of view the use of Newton - Raphson method in the iterative improvement of the solution is not specific. The implementation of stepwise processes is also possible with the help of other iterative procedures. For example the replacement of the matrix $J\left(x_{(i)}^{(j-1)}, p_i\right)$ in (1.5) by $J\left(x_{(0)}^{(j-1)}, p_i\right)$ corresponds to modified Newton method, etc.

Stepwise processes with iterative improvement of the solution at each step will be termed discrete continuation of the solution.

Another formulation of the continuation method was given by D. Davidenko [24, 25] (1953). He was apparently the first who realized the process of solution continuation as a process of moving, and applied adequate mathematical apparatus of differential equations to it. Differentiating equations (1.1) with respect to a parameter p and taking into account initial solution $x_{(0)} = x(p_0)$ he formulated the problem of finding a set of system (1.1) solutions as a Cauchy problem for the system of ordinary differential equation

$$J\frac{dx}{dp} + \frac{\partial F}{\partial p} = 0, \qquad J = \frac{\partial F}{\partial x}, \qquad x(p_0) = x_{(0)}. \qquad (1.6)$$

The system of the equation of this initial - value problem will be termed continuation equation in implicit form.

For this system the equation (1.1) is a complete integral which satisfies the condition $F(x_{(0)}, p_0) = 0$.

System (1.6) is linear with respect to the derivatives dx/dp. According to well - known existence and uniqueness theorems (see [92] as example) the solution of the problem (1.6) exist and is unique in the whole interval of its argument ,i.e. parameter p, in which the Jacobian $\det(J)$ differs from zero. Moreover, the solution is a continuous and differentiable vector function. It is a curve K of the set of solutions of equation (1.1) in the space $^{n+1}$: $\{x,\ p\}$ passing through the point $(x_{(0)}, p_0)$.

In the interval of the parameter p where $\det(J) \neq 0$ the system of ODE (1.6) can be reduced to the normal Cauchy form for which the corresponding initial - value problem becomes

$$\frac{dx}{dp} = -J^{-1}\frac{\partial F}{\partial p}, \qquad x(p_0) = x_{(0)}. \tag{1.7}$$

The system of equation for this problem will be termed as the normal form of continuation equations.

To construct the solution $x(p)$ this approach opens up the possibility of using various well - studied integration schemes for initial value problems. The simplest of these schemes, namely the scheme of Euler's method, leads to the following algorithm:

$$x_{(0)} = x_0; \quad x_{(i+1)} = x_{(i)} - \frac{J^{-1}(x_{(i)},\ p_i)F_{,p}(x_{(i)}, p_i)\Delta p,}{i = 0,\ k-1,} \tag{1.8}$$

where $\Delta p = p_{i+1} = p_i$, $\quad F_{,p} = \partial F/\partial p$.

It is not difficult to construct the algorithms for other schemes which have a higher order of accuracy, such as the modified Euler method, the Runge - Kutta method, the Adams - Stormer method, etc. These schemes were used and studied within the frame work of the continuation method in [25, 26, 60, 61, 10] and in other works.

Let us show that stepwise processes with iterative improvement of the solution similar to Lahaye process (1.5) can also be related to the schemes of integration of the initial value problem for continuation equations (1.7). To do this the solution of this initial value problem at each p - step is represented as

$$x_{(i+1)} = x_{(i)} - \int\limits_{p_i}^{p_{i+1}} J^{-1}(x(p),\ p)F_{,p}(x(p), p)dp. \tag{1.9}$$

Putting $p_{i+1} = p_i + \Delta p$, by the mean value theorem, we obtain from (1.9)

$$x_{(i+1)} = x_{(i)} - J^{-1}(x(p),\ p)F_{,p}(x(p),p)\Delta p, \quad p_i \leq p \leq p_{i+1}. \quad (1.10)$$

This relation may by considered as a nonlinear implicit finite differences scheme which involves a new unknown p, and therefore it must be solved in conjunction with the original equation (1.1). This makes its direct implementation difficult. It is possible to obtain the explicit Euler difference scheme (1.45) which is based on the approximate representation of expression (1.10). Other methods for constructing explicit difference schemes, which are based on various formulas of numerical integration for relation (1.9), are considered ,for example, by Bachvalov [7].

Let us set $J(x(p),\ p) \approx J(x_{(i)},\ p_{i+1})$ in expression (1.10) and the use the following numerical differentiation formula

$$F_{,p}(x(p),p) = \frac{1}{\Delta p}(F(x_{(i)}, p_{i+1}) - F(x_{(i)}, p_i)) + O(\Delta p^2). \quad (1.11)$$

Noting that $F(x_{(i)}, p_i) = 0$, from (1.10) we obtain a relation which is identical to one step of the Newton - Raphson method

$$x_{(i+1)} = x_{(i)} - J^{-1}(x_{(i)},\ p_{i+1})F(x_{(i)}, p_{i+1}) + O(\Delta p^2). \quad (1.12)$$

Let us base the algorithm of the stepwise process for the equation $x(p)$ construction on the relation (1.12) and the choice of the solution $x(p_i)$, obtained on the previous step, as the initial approximation. This algorithm is identical to that of Lahaye 1.5).

Thus the algorithms, that were termed in [48] as continuous and discrete continuation, are reduced to the integration of Cauchy problem (1.7) by means of explicit and implicit schemes correspondingly.

It is worth mentioning one other interesting possibility of using the continuation method to design an iterative solution process for the nonlinear equation $H(x) = 0$ (this possibility was established by Gavurin [41]). We construct an equation with a parameter as follows

$$F(x,p) = H(x) - (1 - p)H(x_{(0)}) = 0, \quad p \in [0, 1]. \quad (1.13)$$

Here the parameter p is introduced so that $x_{(0)}$ is a solution of equation (1.13) when $p = 0$, and when $p = 1$ the equation transforms into the original one. If now a new parameter λ is introduced so that

$$1 - p = e^{-\lambda}, \quad \lambda \in [0, \infty), \tag{1.14}$$

than the equation (1.13) becomes

$$F(x, \lambda) = H(x) - e^{-\lambda}H(x_{(0)}) = 0.$$

Differentiating of this equation with respect to λ leads to the following Cauchy problem with respect to the parameter λ

$$\frac{dx}{d\lambda} = -\left(\frac{\partial H}{\partial x}\right)^{-1} H(x), \quad x(0) = x_{(0)}. \tag{1.15}$$

Integrating this problem with respect to λ by Euler's method with the step $\Delta\lambda = 1$ leads to an iterative process

$$x^{(j+1)} = x^{(j)} - \left(\frac{\partial H(x^{(j)})}{\partial x}\right)^{-1} H(x^{(j)}), \quad x^{(0)} = x_{(0)}, \quad j = 0, 1, 2, \ldots \tag{1.16}$$

This process is exactly the same as the iterative process of the Newton - Raphson method for equation $H(x) = 0$ with initial approximation $x^{(0)} = x_{(0)}$.

The continuation method in the form presented here can be extended, practically unchanged, to nonlinear operator equation assuming that $F(x, p)$ is a nonlinear operator with parameter.

Many methods of solutions of applied nonlinear problems can be understood as a special cases of parametric continuation method. For example, in mechanics of solids it is known the method of incremental loading proposed by Vlasov and Petrov in 1959 [91] for the nonlinear Foppl - Karman equations of plates bending. The algorithm of this method is actually the algorithm of Euler's method for integrating of continuation equation with respect to the parameter of loading.

2.   The problem of choosing the continuation parameter. Replacement of the parameter

The above forms of the continuation method presume that in the considered range of the parameter $p_0 \leq p \leq p_k$ the determinant $\det(J)$ of the Jacobi matrix of the system of equation (1.1) is nonzero. The use of the method in the neighborhood of a singular points, where $\det(J) = 0$, requires of a special discussion.

Consider the system (1.1) written out in full

$$F_i(x_1, x_2, \ldots, x_n, \, p) = 0, \quad i = \overline{1, \, n}. \tag{1.17}$$

Differentiating this system of equations with respect to parameter p leads to implicit form of the continuation equations

$$\sum_{j=1}^{n} \frac{\partial F_i}{\partial x_j} \frac{dx_j}{dp} + \frac{\partial F_i}{\partial p} = 0, \qquad i = \overline{1, \, n}. \tag{1.18}$$

This system of equations is linear with respect to the unknowns dx_j/dp, $j = \overline{1, \, n}$. At regular points of the solution set of system (1.17), where $\det(J) \neq 0$, the system (1.18) is solvable.

Along with the Jacobi matrix $J = [\partial F_i/\partial x_j]$, $i, j = \overline{1, \, n}$ we shall consider an augmented Jacobi matrix $\overline{J}$, formed by adding to J on the right the vector $\partial F_i/\partial p = [\partial F_i/\partial p]$, $i = \overline{1, \, n}$

$$\overline{J} = \left[J, \frac{\partial F}{\partial p} \right]. \tag{1.19}$$

By Cramer's rule, the solution of the system (1.18) can then be written as

$$\frac{dx_i}{dp} = (-1)^{n-i} \frac{\det(J_i)}{\det(J)}, \qquad i = \overline{1, \, n}. \tag{1.20}$$

Here $\det(J_i)$ — is a determinant of the matrix obtained from $\overline{J}$ by deleting the i – column.

Note that the matrix J is square of order n and the augmented matrix $\overline{J}$ – is rectangular of dimension $n \times (n + 1)$, i.e. , it is composed of n rows and $n + 1$ columns.

The rank of a matrix A of dimensions $m \times n$ is designated as $rank(A)$. It is determined by the number of linearly independent columns or rows of the matrix. It is convenient to regard the columns of a matrix A as vectors in space ^{m}, and the rows also as vector, but in space ^{n}. The number of linearly independent rows and columns always coincide.

At regular points of the solution set of the system (1.17) $\det(J) \neq 0$, i.e. the columns of the matrix J form an n – dimension basis in ^{n}. The vector $\partial F/\partial p \in {}^n$ is therefore linearly dependent of the columns of the matrix J and its addition to J in forming $\overline{J}$ does not change the rank of the new system matrix now composed of $(n + 1)$ vectors. I.e., at regular points

$$rank(\overline{J}) = rank(J) = n. \tag{1.21}$$

At singular points the situation is different. Here $\det(J) = 0$, therefore $rank(J) = r < n$, i.e., among n columns of the matrix J only r columns are linearly independent. In ^{n} they form an r – dimension basis B^r, which determines an r – dimension subspace $L^r \subset {}^n$. If now the vector $\partial F/\partial p \in L^r$, i.e. it is linearly dependent on the vectors of the basis B^r, then

$$rank(\overline{J}) = rank(J) = r < n. \tag{1.22}$$

If, however, the vector $\partial F/\partial p \notin L^r$, i.e. it is linearly independent of the vectors of the basis B^r, then

$$rank(\overline{J}) = rank(J) + 1 = r + 1. \tag{1.23}$$

A special case takes place when

$$rank(\overline{J}) = n, \qquad rank(J) = n - 1. \tag{1.24}$$

Points where this condition is fulfilled are usually called limit points. Singular points at which $rank(\overline{J}) < n$ will be termed essentially singular. Peculiarity of the solution behavior at these points will be regarded later. Here we consider only limit points.

It will be shown below that at limit point the tangent to the curve K of the solution set of system (1.17) in ${}^{n+1} : \{x_1, x_2, \ldots, x_n, p\}$ becomes normal to the p axis. At these points the condition $rank(\overline{J}) = n$ is equivalent to the requirement that among $(n+1)$ columns of the matrix $\overline{J}$ there are n linearly independent columns. This means that at least one of the determinants $\det(J_k)$, $k = \overline{1, n}$ is not equal to zero. Let

$$\det(J_j) \neq 0. \tag{1.25}$$

In this case x_j can be taken as a continuation parameter. Then in the vicinity of the limit point all condition of the implicit function theorem are fulfilled. Differentiating the system (1.17) with respect to x_j and assuming that all x_i, $i = \overline{1, n}$, $i \neq j$ and parameter p are functions of x_j we obtain the continuation equations in the form

$$\sum_{l=1}^{n} \frac{\partial F_i}{\partial x_l} \frac{dx_l}{dx_j} + \frac{\partial F_i}{\partial p} \frac{dp}{dx_j} = 0, \qquad i = \overline{1, n}. \tag{1.26}$$

By Cramer's rule we have from (1.26)

$$\frac{dx_i}{dx_j} = (-1)^{j-i} \frac{\det(J_i)}{\det(J_j)},$$

$$\frac{dp}{dx_j} = (-1)^{n-j} \frac{\det(J)}{\det(J_j)}, \tag{1.27}$$

$$i = \overline{1, n}, \quad i \neq j.$$

The fact that $\det(J_j) \neq 0$ removes the difficulties related to the unlimited growth of the solution. Taking into account that $\det(J) = 0$, the last equation in (1.27) shows that at the limit point the tangent to the solution curve K is normal to the p axis. Indeed, we represent the tangent to K as a vector in $^{n+1}$

$$\frac{dx}{dx_j} = \left(\frac{dx_1}{dx_j}, \frac{dx_2}{dx_j}, \ldots, \frac{dx_n}{dx_j}, \frac{dp}{dx_j} \right)^T \in {}^{n+1}.$$

But since $\det(J) = 0$, it follows that $dp/dx_j = 0$. Thus the scalar product of the vector dx/dx_j and the vector of the p axis $(0, 0, \ldots, 0, p)$ is zero.

The transition from the continuation equation with the parameter p (1.20) to the continuation equations with the parameter x_j (1.27) in the neighborhood of a limit points are called the parameter replacement. This method was proposed by Davidenko [25] when he formulated the parametric continuation method.

3. The best continuation parameter

The method of parameter replacement, described in the previous section, shows that from the standpoint of the solution continuation there is no principal difference between the unknowns x_i and the parameter p. Taking this into account we denote $p = x_{n+1}$ and write the problem (1.17) in the form

$$F_i(x_1, x_2, \ldots, x_{n+1}) = 0, \qquad i = \overline{1, n}, \qquad (1.28)$$

or

$$F(x) = 0, \quad x \in {}^{n+1}, \quad F : {}^{n+1} \to {}^n. \qquad (1.29)$$

As follows from the previous section any unknown x_j, $j = \overline{1, n+1}$ for which $\det(J_j) \neq 0$ can be selected as a continuation parameter. Such an ambiguity brings us to the question: is there the best (at least in some sense) continuation parameter?

The problem can be illustrated by the simplest case of one equation with two unknowns

$$F(x_1, x_2) = 0. \qquad (1.30)$$

Let the set of solution s of this equation forms the curve K shown on Fig. 1.2. If the solution continuation process is realized as a process of Cauchy's problem integration with respect to a parameter then it is reduced to some stepwise process that use increments at each step. From this point of view it is obvious that if x_1 is selected as the continuation parameter then the numerical situation will be the best in a vicinity of the point A, because at the point the increment of the argument Δx_1 is much more than increment Δx_2. By approaching to the point B the numerical situation become worse because near this point $\Delta x_2 \gg \Delta x_1$, i.e. , sizable function increment Δx_2 corresponds to a small argument increment Δx_1. And this situation is a typical sign at an unstability.

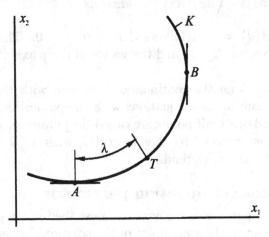

Figure 1.2.

If x_2 is selected in the capacity of the parameter then on the contrary the best numerical situation will be near the point B and an unstability will appear near the point A.

Call attention to the fact that the best situation near points A and B is realized when parameter axis is parallel to the tangent to the curve K in these points. It suggest an idea that if we want to ensure the best numerical situation at the each point of the curve K then it can be attained by choosing the length λ measured along the curve K in capacity of the parameter as it is shown on the Fig. 1.2 for point T. Indeed, locally, i.e. , near the each point of the curve K, this parameter coincide in direction with the tangent to the curve. I. Vorovich and V. Zipalova [121] and E. Riks [98] - [101] were apparently first who offered the use of the arc length as a continuation perameter.

Introducing the λ as a continuation parameter is based on the assumption that unknowns x_1 and x_2 are functions of λ, i.e.

$$x_1 = x_1(\lambda), \qquad x_2 = x_2(\lambda). \tag{1.31}$$

In fact it is equivalent to introducing a new unknown λ that does not be contained explicitly in equation (1.30). To determine this parameter it is necessary to formulate an additional relation between x_1, x_2 and λ. Locally this relation is evident

$$d\lambda^2 = dx_1^2 + dx_2^2. \tag{1.32}$$

Thus the introducing of the parameter λ calls the joint solution of the equation (1.30) and the relation (1.32). As soon the relation (1.32) is differential then it would be logical to use the differential form of the equation (1.30) having applied it's equivalent — the Cauchy problem with respect to the parameter λ. Having differentiating the equation (1.30) with respect to λ and taking into account the known values x_{10}, x_{20}, we obtain the following Cauchy parametric problem

$$\frac{\partial F}{\partial x_l}\frac{dx_l}{d\lambda} + \frac{\partial F}{\partial x_2}\frac{dx_2}{d\lambda} = 0,$$

$$\left(\frac{dx_l}{d\lambda}\right)^2 + \left(\frac{dx_2}{d\lambda}\right)^2 = 1, \tag{1.33}$$

$$x_1(\lambda_0) = x_{10}, \quad x_2(\lambda_0) = x_{20}.$$

It is assumed here that the value λ_0 of the parameter λ corresponds to the point x_{10}, x_{20}.

Without getting into the details of solving the initial - value problem we note that all the arguments mentioned above are most likely heuristic and can not be considered as a strict proof.

In the general case we will suppose that in equations (1.28), (1.29) the unknowns depend of some parameter μ

$$x = x(\mu). \tag{1.34}$$

Then after differentiating equations (1.29) with respect to μ we obtain the continuation equations

$$\overline{J}\frac{dx}{d\mu} = 0, \qquad \overline{J} = \overline{J}(x) = \frac{\partial F}{\partial x}. \tag{1.35}$$

Here $\overline{J}$ coincides with the augmented Jacobi matrix introduced in the previous section. In the vicinity of the point x on the curve of the solutions set we introduce parameter μ that is measured along the axis determined by identify vector $\alpha = (\alpha_1, \ldots, \alpha_{n+1})^T \in {}^{n+1}$, $\alpha\alpha = \alpha_i\alpha_i = 1$. Then in this point

$$d\mu = \alpha dx = \alpha_i dx_i, \quad i = \overline{1, n+1}. \tag{1.36}$$

Here and below the summation with respect to repeating indexes in mentioned limits is used.

Choosing the vector α in a different way we can define any continuation parameter. For example, choosing $\alpha = (1, 0, \ldots, 0)^T$ we obtain from (1.36) that $d\mu = dx_1$ and $\mu = x_1 + C$. And if the arbitrary constant C is assumed equal to zero then $\mu = x_1$. Similarly, if $\alpha = (0, \ldots, 0, 1)^T$ then $\mu = x_{n+1} = p$ etc.

The continuation equations (1.35), (1.36) we write as

$$
\overline{J}\frac{dx}{d\mu} = \begin{bmatrix} J^{(1)} \\ J^{(2)} \\ \ldots \\ J^{(n)} \end{bmatrix} \frac{dx}{d\mu} = 0,
$$

$$
\alpha\frac{dx}{d\mu} = 1,
$$

$$
J^{(i)} \in {}^{n+1}, \quad i = \overline{1, n}.
$$

(1.37)

Here the matrix $\overline{J}$ is represented as an assemblage of its rows $J^{(i)}$ and each of them can be regarded as a vector in ${}^{n+1}$. At least one of the determinants $det(J_j) \neq 0$ in a regular and limit points of the set of solution of the equations (1.29). Therefore the rows of the matrix $\overline{J}$ are linearly independent and form n – dimensional basis B^n that determines n – dimensional subspace $L^n \subset {}^{n+1}$.

The first vector equation or the first n scalar equations (1.37) show that desired vector $dx/d\mu$ is orthogonal to all vectors of the basis B^n, i.e., it is orthogonal to the subspace L^n. Geometrically, it points in the direction of the tangent of the solutions curve K. The last equation in (1.37) shows that its projection on the vector α, which determines the axis μ, must be equal one. It is shown geometrically in Fig. 1.3 for the case $n = 1$.

It becomes obvious now that if vector α (axis μ) is choosed orthogonal to the curve K, as it shown by dotted line at Fig. 1.3, then the square matrix $\begin{bmatrix} \overline{J} \\ \alpha \end{bmatrix}$ of the system (1.37) becomes singular since the vector α is orthogonal to $dx/d\mu$ so it belongs to the subspace L^n, i.e. , it is linearly dependent with the rows of the matrix $\overline{J}$. As far as for integration of continuation equations (1.37) it is necessary to pass to the normal form of equations, i.e., to resolve them with respect to $dx/d\mu$, so this choice of the parameter μ is inadmissible.

And, on the contrary, it should be expected that the most successful will be such a choice of the axis μ when the vector α is tangent to the curve K, i.e., is collinear to $dx/d\mu$.

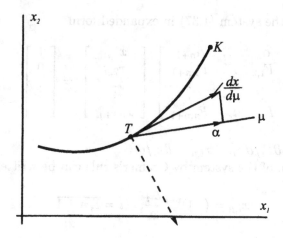

Figure 1.3.

To attach the sufficient strictness to these arguments it is left only to introduce some criterion evaluating the quality of the parameter μ defined by the vector α. As soon as the quality of the parameter μ is related with decidability of the system (1.37) then it is naturally to connect the quality with the conditionality of the matrix of this system. The smaller are the variations of the system solutions caused by small variations of the matrix elements or right – hand part of the system the better conditioned the matrix of the system of the linear algebraic equation is considered.

As the measure of the conditionality $|D|$ of the matrix

$$A = [a_{ij}] = \begin{bmatrix} t_1 \\ t_2 \\ \dots \\ t_n \end{bmatrix}, \qquad i, j = \overline{1, n},$$

which has vectors t_i, $i = \overline{1, n}$, as its rows, we take the module of its determinant $\det(A) = \Delta$ divided by product squared measures of its rows

$$|D| = \frac{|\Delta|}{\prod_{i=1}^{n}(t_i t_i)^{\frac{1}{2}}}. \tag{1.38}$$

It is shown in [89] that better conditionality corresponds to greater value of $|D|$.

We rewrite the system (1.37) in expanded form

$$
\begin{bmatrix}
\alpha_1 & \alpha_2 & \cdots & \alpha_{n+1} \\
F_{1,1} & F_{1,2} & \cdots & F_{1,n+1} \\
\vdots & \vdots & \ddots & \vdots \\
F_{n,1} & F_{n,2} & \cdots & F_{n,n+1}
\end{bmatrix}
\begin{bmatrix}
x_{1,\mu} \\
x_{2,\mu} \\
\vdots \\
x_{n+1,\mu}
\end{bmatrix}
=
\begin{bmatrix}
1 \\
0 \\
\vdots \\
0
\end{bmatrix}.
\tag{1.39}
$$

Here $F_{i,j} = \partial F_i / \partial x_j$, $x_{i,\mu} = dx_i / d\mu$.

The solution of the system by Cramer's rule can be written as

$$
x_{i,\mu} = (-1)^{i+1} \frac{\Delta_i}{\Delta}, \quad i = \overline{1, n+1}
\tag{1.40}
$$

Here Δ_i is the determinant obtained from the augmented Jacoby matrix $\overline{J}$ by deletion of i - th column. At the same time it is a cofactor of α_i in the matrix of the system (1.39). If we uncover the determinant Δ of this system with respect to the elements of the first row then

$$
\Delta = (-1)^{i+1} \alpha_i \Delta_i, \quad i = \overline{1, \, n+1}.
\tag{1.41}
$$

Let us prove that the following statement is valid [108, 66, 67].

Lemma 1. The conditionality measure $|D|$ of the matrix of the equations system (1.39) attains its greatest value when the vector α is tangent to the curve K of the set of the solutions system (1.28) in the considered point.

PROOF. Let us investigate the extremum of the value D as a function of α_i, $i = \overline{1, \, n+1}$ (1.38) which can be presented in the form

$$
D(\alpha_1, \ldots, \alpha_{n+1}) = \frac{\Delta}{d},
\tag{1.42}
$$

where $d = d_1 d_2 \ldots d_{n+1} > 0$;

$$
d_1 = (\alpha_i \alpha_i)^{\frac{1}{2}} = 1, \quad d_{\beta+1} = (F_{\beta,i} F_{\beta,i})^{\frac{1}{2}}, \quad i = \overline{1, \, n+1}, \quad \beta = \overline{1, \, n}.
$$

Here the summation is not taken over index β.

Let us note that since α is a unit vector then $d_1 = 1$ and thus the value of d does not depend on α_i, $i = \overline{1, \, n+1}$.

To find an extremum of the function D provided that $\alpha\alpha = 1$ we compose the Lagrange's function with regard the expression (1.41) for Δ

$$
L = (-1)^{i+1} \alpha_i \frac{\Delta_i}{d} + \gamma(1 - \alpha_i \alpha_i), \quad i = \overline{1, \, n+1},
\tag{1.43}
$$

Where γ is indeterminate Lagrange multiplier.

Extremum of the function is reached when

$$\frac{\partial L}{\partial \alpha_k} = (-1)^{k+1}\frac{\Delta_k}{d} - 2\gamma\alpha_k = 0, \qquad k = \overline{1,\, n+1},$$

i.e., when $\alpha_k = (-1)^{k+1}\frac{\Delta_k}{2\gamma d}$. Substituting this value of α_k to the condition $\alpha\alpha = 1$ we find the Lagrange multiplier

$$\gamma = \pm\frac{(\Delta_i\Delta_i)^{\frac{1}{2}}}{2d}. \tag{1.44}$$

Thus the extremum of the Lagrange's function is reached when

$$\alpha_k = \pm(-1)^{k+1}\frac{\Delta_k}{(\Delta_i\Delta_i)^{\frac{1}{2}}}, \qquad k = \overline{1,\, n+1}. \tag{1.45}$$

After substituting the expression of α_k into equality (1.41) we obtain that the determinant of the system (1.39) must satisfy the equality

$$\Delta = \pm(\Delta_i\Delta_i)^{\frac{1}{2}}, \qquad i = \overline{1,\, n+1} \tag{1.46}$$

and the extremum of the Lagrange's function is reached when $\alpha_k = (-1)^{k+1}\frac{\Delta_k}{\Delta}$ which is exactly the same as the expression for $x_{k,\mu}$ (1.40). Thus, the following equalities take place

$$\alpha_k = (-1)^{k+1}\frac{\Delta_k}{\Delta} = x_{k,\mu}. \tag{1.47}$$

Then according to (1.44), (1.46) the value of the Lagrange multiplier is

$$\gamma = \frac{\Delta}{2d} = \frac{D}{2} \tag{1.48}$$

Analysis of the second differential of the Lagrange's function which is a quadratic form with respect to differentials $d\alpha_k$ shows that the modulus of function $D = \frac{\Delta}{d}$ reaches in this case its maximal value $|D| = \frac{(\Delta_i\Delta_i)^{\frac{1}{2}}}{d}$. Indeed, the sign of the Lagrange's function second differential

$$d^2L = -2\gamma(d\alpha_i d\alpha_i)$$

is determined by the Lagrange multiplier γ, which according to formula (1.48) is positive if $D > 0$. Therefore, the function D reaches its maximal value. The sign of the multiplier γ is negative if $D < 0$ and, therefore, function D takes the minimal value. The lemma is proved.

Thus, it is proved that the vector α, which determines in according to expression (1.36) the continuation parameter μ, provides the maximal value of the modulus of the of the function D in the case when it coincides with the solution vector $(x_{1,\mu}, x_{2,\mu}, \ldots, x_{n+1,\mu})^T$ of the linearized system (1.37), i.e., when the vector α is pointed along the tangent to the curve of the solutions set of the nonlinear equation system (1.28).

Let us consider now an influence of perturbations of the elements of the system (1.39) matrix on its conditionality. We will prove that the following statement takes place.

Lemma 2. Squared error of the solution of the system (1.39) arising from the perturbation of the system matrix elements reaches its minimal value when the vector α is pointed along the tangent to the curve of solution set of the system (1.28) in the considered point.

PROOF. At the beginning we consider a case when the first row of the system (1.39) matrix is given with the error. Let the vector α has the form $(\alpha_1, \alpha_2, \ldots, \alpha_{j-1}, \alpha_j + \varepsilon, \alpha_{j+1}, \ldots, \alpha_{n+1})^T$. The determinant Δ_e of this system is expressed in terms of determinant Δ of the original system as

$$\Delta_e = \Delta + (-1)^{j+1}\varepsilon\Delta_j = \Delta\left(1 + (-1)^{j+1}\varepsilon\frac{\Delta_j}{\Delta}\right).$$

Since small perturbations are considered then the components of the perturbed solution $y_{i,\mu}$ can be written as follows

$$y_{i,\mu} = (-1)^{i+1}\frac{\Delta_i}{\Delta_\varepsilon} = (-1)^{i+1}\frac{\Delta_i}{\Delta}\left(1 - (-1)^{j+1}\varepsilon\frac{\Delta_j}{\Delta}\right).$$

Then components δ_i of the error vector $\delta = (\delta_1, \delta_2, \ldots, \delta_{n+1})^T$ of the perturbed system solution are calculated as

$$\delta_i = y_{i,\mu} - x_{i,\mu} = (-1)^{i+j+1}\varepsilon\frac{\Delta_i\Delta_j}{\Delta^2}, \qquad i = \overline{1,\, n+1}.$$

Let us investigate an extremum of the squared error $\delta^2 = \varepsilon^2\frac{\Delta_j^2\Delta_i\Delta_i}{\Delta^4}$ assuming that the condition $\alpha\alpha = 1$ takes place. Then the Lagrange function can be taken as

$$L = \varepsilon^2\frac{\Delta_j^2\Delta_i\Delta_i}{\Delta^4} + \gamma(\alpha_i\alpha_i - 1), \qquad i = \overline{1,\, n+1}.$$

The minimum of this function is reached when

$$\frac{\partial L}{\partial \alpha_k} = \varepsilon^2\frac{\Delta_j^2\Delta_i\Delta_i}{\Delta^5}(-4)(-1)^{k+1}\Delta_k + 2\gamma\alpha_k = 0.$$

From this equation we obtain

$$\alpha_k = 2\varepsilon^2(-1)^{k+1}\frac{\Delta_j^2\Delta_i\Delta_i\Delta_k}{\gamma\Delta^5}, \qquad k = \overline{1, \, n+1}. \qquad (1.49)$$

Dividing the k – th equation of these relation by the m – th one we obtain the equality

$$\frac{\alpha_k}{\alpha_m} = (-1)^{k-m}\frac{\Delta_k}{\Delta_m}, \qquad m = \overline{1, \, n+1},$$

permitting to express α_m through α_k

$$\alpha_m = (-1)^{m-k}\alpha_k\frac{\Delta_m}{\Delta_k}. \qquad (1.50)$$

Then the determinant (1.41) of the system (1.39) can be represented as

$$\Delta = (-1)^{m-1}\alpha_m\Delta_m = (-1)^{-k-1}\alpha_k\frac{\Delta_m\Delta_m}{\Delta_k}, \qquad m = \overline{1, \, n+1}. \quad (1.51)$$

In this case the equations system (1.49) takes the form

$$\alpha_k = \frac{2\varepsilon^2\Delta_j^2}{\gamma\alpha_k^5(\Delta_m\Delta_m)^4}(-1)^{6k+6}\Delta_k^6, \qquad m = \overline{1, \, n+1} \qquad (1.52)$$

and can be resolved easily with respect to α_k

$$\alpha_k = \left(\frac{2\varepsilon^2\Delta_j^2}{\gamma(\Delta_m\Delta_m)^4}\right)^{\frac{1}{6}}(-1)^{k+1}\Delta_k, \qquad m = \overline{1, \, n+1}. \qquad (1.53)$$

Note that there is no summation with respect to index k in expressions (1.50)-(1.52).

Lagrange multiplier γ is found by substituting (1.53) in the condition $\alpha\alpha = 1$. Then $\gamma = 2\varepsilon^2\dfrac{\Delta_j^2}{\Delta_m\Delta_m}$ and expression (1.53) takes the form

$$\alpha_k = (-1)^{k+1}\frac{\Delta_k}{(\Delta_m\Delta_m)^{\frac{1}{2}}}, \qquad m = \overline{1, \, n+1} \qquad (1.54)$$

If now these values of α_k are substituted in the formula (1.41) then we obtain that $\Delta = (\Delta_m\Delta_m)^{\frac{1}{2}}$ and equalities (1.54) are exactly the same as equalities (1.40) for $x_{k,\mu}$, i.e., the relations (1.47) is valid as was to be shown.

Consider now the situation when the row other then the first one is given with error. We develop the proof in the case when $n = 1$. Then the considered nonperturbed system take the form

$$\begin{pmatrix} \alpha_1 & \alpha_2 \\ F_{1,1} & F_{1,2} \end{pmatrix} \begin{pmatrix} x_{1,\mu} \\ x_{2,\mu} \end{pmatrix} = \begin{pmatrix} 1 \\ 0 \end{pmatrix}. \qquad (1.55)$$

Let us assume that the second row of the matrix is given with an error $(F_{1,1}\ F_{1,2} + \varepsilon)$. Then the determinant Δ_e of the perturbed system is expressed in terms of determinant Δ of the initial system (1.55) as

$$\Delta_e = \Delta \left(1 + \frac{\varepsilon \alpha_1}{\Delta}\right).$$

Then perturbed system has the solution

$$y_{1,\mu} = \frac{\Delta_1 + \varepsilon}{\Delta_\varepsilon}, \qquad y_{2,\mu} = -\frac{\Delta_2}{\Delta_\varepsilon},$$

where $\Delta_1 = F_{1,2}$, $\Delta_2 = F_{1,1}$ and taking into account that ε is infinitesimal, the components of the error vector take the form

$$\delta_1 = \frac{\Delta_1\left(1 + \dfrac{\varepsilon}{\Delta_1}\right)}{\Delta\left(1 + \dfrac{\varepsilon\alpha_1}{\Delta}\right)} - \frac{\Delta_1}{\Delta} =$$

$$= \frac{\Delta_1}{\Delta}\left[\left(1 + \frac{\varepsilon}{\Delta_1}\right)\left(1 - \frac{\varepsilon\alpha_1}{\Delta}\right) - 1\right] = -\frac{\alpha_2\Delta_2}{\Delta^2}\varepsilon$$

$$\delta_2 = -\frac{\Delta_2}{\Delta\left(1 + \dfrac{\varepsilon\alpha_1}{\Delta}\right)} + \frac{\Delta_2}{\Delta} = \frac{\alpha_1\Delta_2}{\Delta^2}\varepsilon.$$

To test the square error $\delta^2 = \dfrac{\varepsilon^2\Delta_2^2}{\Delta^4}$ for an extremum by the condition $\alpha^2 = 1$ we draw up the Lagrange function

$$L = \varepsilon^2\frac{\Delta_2^2}{\Delta^4} + \gamma(\alpha_i\alpha_i - 1), \qquad i = 1, 2.$$

If the previously used algorithm is applied for searching the function for an extremum is employed then we obtain that the vector components must satisfy the equalities (1.47). So the lemma is proved for $n = 1$.

Eventually we consider the general case when an arbitrary column of the matrix of the system (1.39) is given with perturbation.   In this case we perturb the system matrix elements by adding the vector $\varepsilon(\alpha_j, F_{1,j}, \ldots, F_{n,j})^T$ to the j – th column. Here ε – is a small value. Then it is easy to see that the determinants of the perturbed $(\Delta_e, \Delta_{i\varepsilon})$ and nonperturbed (Δ, Δ_i) systems are connected by relations $\Delta_e = \Delta(1+\varepsilon)$, $\Delta_{i\varepsilon} = \Delta_i(1+\varepsilon)$, $i = \overline{1, n+1}, i \neq j$, $\Delta_{j\varepsilon} = \Delta_j$.

The solution of the perturbed system takes the form

$$y_{i,\mu} = (-1)^{i+1}\frac{\Delta_i}{\Delta}, \qquad i = \overline{1, n+1}, \qquad i \neq j,$$

$$y_{j,\mu} = (-1)^{j+1}\frac{\Delta_j}{\Delta(1+\varepsilon)}.$$

Components of the error vector $\delta = (\delta_1, \delta_2, \ldots, \delta_{n+1})^T$ counted by formulas $\delta_i = y_{i,\mu} - x_{i,\mu}$ take the values $\delta_i = 0$, $i = \overline{1, n+1}$, $i \neq j$, $\delta_j = (-1)^j \varepsilon \frac{\Delta_j}{\Delta(1+\varepsilon)}$.

Thus, the squared error $\delta^2 = \delta_i\delta_i$ is $\delta^2 = \varepsilon^2\dfrac{\Delta_j^2}{(\Delta(1+\varepsilon))^2}$ and by researching it for an extremum when the vector α is unit the Lagrange function can be given as

$$L = \varepsilon^2\frac{\Delta_j^2}{(\Delta(1+\varepsilon))^2} + \gamma(\quad - 1), \qquad i = \overline{1, n+1}.$$

If now we use the previously employed algorithm for searching an extremum of the Lagrange function, then the proof of the lemma 2 will be obtained.

Let us study the influence of the right - hand side of the system (1.39) on its conditionality.

We will show now that the following statement is correct:

Lemma 3. The squared error of the solution of the system (1.39) occurring due to the perturbations of the right - hand side of the system reaches its minimal value when the vector α is tangential to the curve of the system (1.28) solution set in the point considered.

PROOF. First consider the case when the vector of the perturbed right - hand side of the system (1.39) takes the form $(1+\varepsilon, 0, \ldots, 0)^T$. Then the perturbation vector is $\delta = \varepsilon(x_{1,\mu}, x_{2,\mu}, \ldots, x_{n+1,\mu})^T$ and squared error is

$$\delta^2 = \varepsilon^2\frac{\Delta_i\Delta_i}{\Delta^2}, \qquad i = \overline{1, n+1}.$$

Lagrange function can be given as

$$L = \varepsilon^2\frac{\Delta_i\Delta_i}{\Delta^2} + \gamma(\alpha_i\alpha_i - 1), \qquad i = \overline{1, n+1}$$

If it is searched for an extremum by means of the method described previously then we obtain that in the point of extremum the vector components must satisfy the equalities (1.47). So the lemma is proved for this case.

We prove the other case for the system (1.55) as an example. Let the vector of the perturbed right - hand side of this system to take the form $(1, \varepsilon)^T$. Then the solution of the perturbed equation is

$$y_{1,\mu} = \frac{(\Delta_1 - \varepsilon\alpha_2)}{\Delta}, \qquad y_{2,\mu} = -\frac{(\Delta_2 - \varepsilon\alpha_1)}{\Delta}$$

and the error vector takes the form

$$\delta = \frac{(-\alpha_2,\ \alpha_1)^T \varepsilon}{\Delta}.$$

Taking into account that the condition $\alpha^2 = 1$ the squared error can be written as $\delta^2 = \varepsilon^2/\Delta^2$ results in the Lagrange function $L = \varepsilon^2/\Delta^2 + \gamma(\alpha_i\alpha_i - 1)$, $i = 1, 2$. Searching this function for an extremum we obtain that in the minimum point the components of the vector α must be found by formulas (1.47).

Thus, the lemma is proved for $n = 1$. To prove the general case we perturb the column of the system (1.39) right - hand side adding the vector $\varepsilon(1 + \alpha_1, F_{1,1}, \ \ldots, F_{n,1})^T$ to it .

Taking into account the properties of determinants it is easy to demonstrate that in this case the determinants of perturbed and nonperturbed systems are related as follows

$$\Delta_e = \Delta, \ \Delta_{1\varepsilon} = \Delta_1 + \varepsilon(\Delta_1 + \Delta),$$

$$\Delta_{j\varepsilon} = \Delta_j(1 + \varepsilon), \qquad j = \overline{2,\ n+1}.$$

The solution of the perturbed system and the components of the error vector can be represented as

$$y_{j,\mu} = (-1)^{j+1}\frac{\Delta_{j\varepsilon}}{\Delta},$$

$$\delta_1 = \varepsilon\left(1 + \frac{\Delta_1}{\Delta}\right), \ \delta_j = (-1)^{j+1}\varepsilon\frac{\Delta_j}{\Delta}, \qquad j = \overline{2, n+1}.$$

Therefore the squared error is

$$\delta^2 = \varepsilon^2\left(1 + 2\frac{\Delta_1}{\Delta} + \frac{\Delta_j\Delta_j}{\Delta^2}\right), \qquad j = \overline{2, n+1}.$$

To search it for an extremum one can take into account that α is an unit vector and use $L = \delta^2 + \gamma(\alpha_j\alpha_j - 1)$ as the Lagrange function. The minimum of this function is reached when

$$\frac{\partial L}{\partial \alpha_k} = -2\varepsilon^2(-1)^{k+1}\Delta_k\left(\frac{\Delta_1}{\Delta^2} + \frac{\Delta_j\Delta_j}{\Delta^3}\right) + 2\gamma\alpha_k = 0,$$

$$j,k = \overline{1, n+1}.$$

Hence, in the point of the minimum

$$\alpha_k = (-1)^{k+1}\omega\Delta_k, \qquad \omega = \frac{\varepsilon^2\left(\frac{\Delta_1}{\Delta^2} + \frac{\Delta_j\Delta_j}{\Delta^3}\right)}{\gamma}. \qquad (1.56)$$

Then it follows from the condition $\alpha^2 = 1$ that

$$\omega = \pm(\Delta_k\Delta_k)^{-\frac{1}{2}}, \qquad k = \overline{1, n+1}. \qquad (1.57)$$

If the relation (1.56) is substituted into expression (1.41) then with the help of (1.57) we obtain

$$\Delta = \pm(\Delta_k\Delta_k)^{\frac{1}{2}},$$

It means that the equality $\omega = 1/\Delta$ is correct and the formulas (1.56) take the form

$$\alpha_k = (-1)^{k+1}\frac{\Delta_k}{\Delta}, \qquad k = \overline{1, n+1}.$$

Comparing these relations for α_k with expressions (1.40) we obtain

$$\alpha_k = x_{k,\mu},$$

which mean that the minimal squared error is provided by the continuation parameter that is measured along the curve of the equation (1.28) solution set. This proves the lemma in general case.

Eventually we can prove now the main result of the chapter

Theorem 1. For the system of linearized equations (1.39) to become the best conditioned, it is necessary and sufficient to take the length of the curve of the system solution set as the continuation parameter of the nonlinear system (1.28)

PROOF. NECESSITY. According to the meaning of the conditionality the proofed lemma can be joined in the following statement.

The system of linear equations (1.39) is the best conditioned system when the vector α is tangential to the curve of the solution set of the nonlinear equation system (1.28) at the each point, i.e., when the equalities (1.47) take place. According to this the expression $\alpha^2 = 1$ can be written as

$$(d\mu)^2 = dx_i dx_i, \qquad i = \overline{1, \, n+1}, \tag{1.58}$$

It follows from the above expression that $d\mu = (dx_i dx_i)^{\frac{1}{2}}$ is the differential of the length of the solution set curve of the system (1.28). If we assume that $\mu = 0$ corresponds to the initial point of the curve, then the continuation parameter μ equals to the length of the curve measured from this point. The necessity has been proved.

SUFFICIENCY. Let us take the length of the solution set curve of the system (1.28) as the continuation parameter μ. The vector τ that is tangential to this curve is $\tau = (x_{1,\mu}, \, \ldots \, , \, x_{n+1,\mu})^T$. As it was noted previously the meaning of the vector α is that it determines the local direction of the continuation parameter of the system (1.28) solution. Therefore, on account of the chosen continuation parameter it must be pointed along the tangent to the solution set curve, i.e., the vectors α and τ must be collinear. But they are not only collinear. They are equal because the vector τ is unity vector too. Indeed, being the curve length element, the differential of the chosen parameter must satisfy the equality (1.58). Divided by $(d\mu)^2$ this equality gives

$$x_{i,\mu} x_{i\mu} = \tau^2 = 1, \qquad i = \overline{1, \, n+1}.$$

Since vectors are equal then their components are equal too

$$\alpha_k = \frac{dx_k}{d\mu}, \qquad k = \overline{1, \, n+1}. \tag{1.59}$$

But the components $dx_k/d\mu = x_{k,\mu}$ must satisfy the linear equation system (1.39) for an arbitrary continuation parameter μ , i.e., they must be determined according to the formulas (1.40). The equalities (1.47) follows from the relations (1.59), (1.40). The left - hand sides of the equalities ensure the corresponding extremes of the functions that appear in the considered lemmas. The theorem is proved.

More general results for smooth Banach manifolds were stated in [116].

4.   The algorithms using the best continuation parameter and examples of their application

It is shown above that the solution of the Cauchy problem of the ordinary differential continuation equations system corresponds to the process of the solution continuation of the nonlinear equation system

$$F(x) = 0, \quad F : {}^{n+1} \to {}^n, \quad x \in {}^{n+1}, \quad F(x_0) = 0$$

when the vector α is tangential to the solution set curve K at each point of the curve, i.e., the vector α coincides with the vector $x_\lambda = dx/d\lambda$. It is convenient to write this Cauchy problem as

$$\begin{bmatrix} J \\ x_{,\lambda}^T \end{bmatrix} x_{,\lambda} = \begin{bmatrix} 0 \\ 1 \end{bmatrix}, \qquad x(\lambda_0) = x_0, \qquad x \in {}^{n+1}. \qquad (1.60)$$

Here the matrix J is the Jacoby matrix of the vector function $F(x)$

$$J = \frac{\partial F}{\partial x} = \frac{\partial(F_1, \ldots, F_n)}{\partial(x_1, \ldots, x_{n+1})} = \begin{bmatrix} F_{i,j} = \dfrac{\partial F_i}{\partial x_j} \end{bmatrix}, \quad i = \overline{1, n}, \quad j = \overline{1, n+1},$$

which coincides with the augmented matrix of Jacoby introduced above.

It is shown that for reducing this problem to the normal Cauchy form it is necessary to solve the system of nonlinear equations. An approach to this difficult problem was suggested in [48] within the framework of the step by step process of the solution continuation. It is based on the fact that the last row of the matrix of the equation (1.60) defines the instantaneous position of the vector along which the continuation parameter λ is measured. Changing of the vector $x_{,\lambda}$ by another vector α in the matrix means changing of the best continuation parameter by a parameter μ that is measured along α. The nearer the vector α to the vector $x_{,\lambda}$ the closer will be the parameter μ to the best continuation parameter λ. Let us denote the vector α by $x_{,\lambda}^*$. Then the Cauchy problem for continuation of the solution set curve from k - th point to $(k+1)$ - th point takes the following form

$$\begin{bmatrix} J \\ x_{,\lambda}^{*T} \end{bmatrix} x_{,\lambda} = \begin{bmatrix} 0 \\ 1 \end{bmatrix}, \qquad x(\lambda_k) = x_k, \qquad \lambda_k \le \lambda \le \lambda_{k+1}. \qquad (1.61)$$

The matrix of the system we shall call the augmented Jacoby matrix.

In the following examples of concrete finite differences schemes we give the ways of formation of the vector $x_{,\lambda}^*$.

a. The explicit scheme of Euler method.

Here we take the vector $x_{,\lambda(k-1)}$ counted on the previous step in capacity of the vector $x_{,\lambda}^*$. Then the problem of choice of the vector at initial point $k = 0$ arises. To solve this problem we take into account that for real problems the initial point is not usually a limiting point with respect to the parameter of this problem. So we can take $x_{,\lambda(-1)}^* = (0, \ldots, 0, 1)^T$.

Note that because of the structure of the differential equations system (1.61) the vector $x_{,\lambda}$ obtained on the k - th step of its solution is tangential to the integral curve of the problem. But the last equation of the system $x_{,\lambda}^{*T} x_{,\lambda} = 1$ means that the vector is not unit since its projection on the vector $x_{,\lambda}^{*}$ counted at the previous step is unit. Thus, if we norm the vector $x_{,\lambda}$ to unity then we obtain the solution of the initial equation (1.60) on the k - th step of the step by step process.

Therefore the scheme of the explicit Euler method takes the form

$$x_{(0)} = x(\lambda_0), \qquad x_{,\lambda(-1)}^{*T} = (0, \ \ldots \ , \ 0, \ 1);$$

$$x_{,\lambda(k)} = \begin{bmatrix} J_{(k)} \\ x_{,\lambda(k-1)}^{*T} \end{bmatrix}^{-1} \begin{bmatrix} 0 \\ 1 \end{bmatrix}, \qquad x_{,\lambda(k)}^{*} = \frac{x_{,\lambda(k)}}{\| x_{,\lambda(k)} \|},$$

$$x_{(k+1)} = x_{(k)} + x_{,\lambda(k)}^{*} \Delta\lambda_k, \qquad\qquad (1.62)$$

$$\lambda_{k+1} = \lambda_k + \Delta\lambda_k, \quad k = 0, 1, 2, \ldots$$

Here $\Delta\lambda_k$ is the step of integration with respect to λ, $J_{(k)} = J(x_{(k)})$, $\|x\| = (x, \ x)^{1/2}$ is the Euclidean norm of the vector x.

It is evident that it is not necessary to inverse of the matrix of the system (1.61). Below the notation $x_{,\lambda(k)}$ will mean the result of the solution of the linear algebraic equations by any effective and economical method. As a rule the Gauss method can be used.

b. The explicit scheme of the modified Euler method.

$$x_{(0)} = x(\lambda_0), \qquad x_{,\lambda(-1)}^{*T} = (0, \ \ldots, \ 0, \ 1);$$

$$x_{,\lambda(k)}^{1} = \begin{bmatrix} J_{(k)} \\ x_{,\lambda(k-1)}^{*T} \end{bmatrix}^{-1} \begin{bmatrix} 0 \\ 1 \end{bmatrix}, \qquad x_{,\lambda(k)}^{1*} = \frac{x_{,\lambda(k)}^{1}}{\| x_{,\lambda(k)}^{1} \|},$$

$$x_{(k+1)}^{1} = x_{(k)} + x_{,\lambda(k)}^{1*} \Delta\lambda_k,$$

$$x_{,\lambda(k)}^{2} = \begin{bmatrix} J(x_{(k+1)}^{1}) \\ x_{,\lambda(k)}^{1*T} \end{bmatrix}^{-1} \begin{bmatrix} 0 \\ 1 \end{bmatrix}, \qquad x_{,\lambda(k)}^{2*} = \frac{x_{,\lambda(k)}^{2}}{\| x_{,\lambda(k)}^{2} \|},$$

$$x_{,\lambda(k)} = \frac{1}{2}(x_{,\lambda(k)}^{1*} + x_{,\lambda(k)}^{2*}), \qquad x_{,\lambda(k)}^{*} = \frac{x_{,\lambda(k)}}{\| x_{,\lambda(k)} \|},$$

$$x_{(k+1)} = x_{(k)} + x_{,\lambda(k)}^{*} \Delta\lambda_k, \qquad\qquad (1.63)$$

$$\lambda_{k+1} = \lambda_k + \Delta\lambda_k, \qquad k = 0, 1, 2, \ldots$$

c. The implicit scheme of Euler method.

The implicit scheme of Euler method we realize as the prediction – correction method.

Prediction: $x_{(0)} = x(\lambda_0)$, $x^{*T}_{,\lambda(-1)} = (0, \ldots, 0, 1)$;

$$x^p_{,\lambda(k)} = \begin{bmatrix} J(x_{(k)}) \\ x^{*T}_{,\lambda(k-1)} \end{bmatrix}^{-1} \begin{bmatrix} 0 \\ 1 \end{bmatrix}, \qquad x^{p*}_{,\lambda(k)} = \frac{x^p_{,\lambda(k)}}{\| x^p_{,\lambda(k)} \|},$$

$$x^p_{(k+1)} = x_{(k)} + x^{p*}_{,\lambda(k)} \, \Delta\lambda_k.$$

Correction:

$$x^c_{,\lambda(k)} = \begin{bmatrix} J(x^p_{(k+1)}) \\ x^{p*T}_{,\lambda(k)} \end{bmatrix}^{-1} \begin{bmatrix} 0 \\ 1 \end{bmatrix}, \qquad x^{c*}_{,\lambda(k)} = \frac{x^c_{,\lambda(k)}}{\| x^c_{,\lambda(k)} \|},$$

$$x^c_{(k+1)} = x_{(k)} + x^{c*}_{,\lambda(k)} \, \Delta\lambda_k, \tag{1.64}$$

$$\lambda_{k+1} = \lambda_k + \Delta\lambda_k, \qquad k = 0, 1, 2, \ldots$$

If the norm of difference between corrected and predicted solutions $\Delta = \| x^c_{(k+1)} - x^p_{(k+1)} \|$ does not satisfy the preassigned accuracy ε then the iteration process described by formulas (1.64) can be used. In this process the value obtained on the previous step of the correction $x^p_{(k+1)} = x^c_{(k+1)}$, $x^{p*}_{,\lambda(k)} = x^{c*}_{,\lambda(k)}$ is taken as a correction. If the conditions of the accuracy are satisfied then at the following step it is necessary to take $x_{(k+1)} = x^c_{(k+1)}$, $x^*_{,\lambda(k)} = x^{c*}_{,\lambda(k)}$

d. Implicit scheme of the second order of accuracy.

Let us consider the two - step method of prediction - correction of the second order of accuracy based on the back differentiation formulas (BDF) [22, 42].

According to recommendation [7, 95] it is expedient to calculate the prediction by extrapolation formulas which use only the values of the solution vector at previous points. First prediction is calculated by formula

$$x^p_{(k+1)} = 2x_{(k)} - x_{(k-1)}.$$

Correction is calculated as

$$x^c_{,\lambda(k)} = \begin{bmatrix} J(x^p_{(k+1)}) \\ x^{*T}_{,\lambda(k-1)} \end{bmatrix}^{-1} \begin{bmatrix} 0 \\ 1 \end{bmatrix}, \qquad x^{c*}_{,\lambda(k)} = \frac{x^c_{,\lambda(k)}}{\| x^c_{,\lambda(k)} \|},$$

$$x^c_{(k+1)} = \frac{1}{3}(4x_{(k)} - x_{(k-1)} + 2x,^{c*}_{\lambda(k)} \Delta\lambda_k). \qquad (1.65)$$

At the first step ($k = 0$) it was taken that $x_{(0)} = x(\lambda_0)$, $\lambda_{k+1} = \lambda_k + \Delta\lambda_k$.

If norm $\Delta \geq \varepsilon$ then the iteration process can be organized with the help of formulas (1.65) where $x^p_{(k+1)} = x^c_{(k+1)}$, $x,^*_{\lambda(k-1)} = x,^{c*}_{\lambda(k)}$. If $\Delta < \varepsilon$ then for next step we have $x_{(k+1)} = x^c_{(k+1)}$, $x,^*_{\lambda(k)} = x,^{c*}_{\lambda(k)}$.

In the book [48] it was suggested also to use the Gram - Schmidt orthogonalization procedure for transformation the continuation equations (1.61) to the Cauchy normal form. This procedure requires also completion of the of the Jacoby matrix J by addition the vector $x,^{*T}_\lambda$ to its low side. But this approach requires more computation then the Gauss method if vectors $x,^{*T}_\lambda$ are defined identically.

Let us consider two illustrative test problems solved in [48].

Bernoully lemniscate.

The equation of the lemniscate in the x_1, x_2 axes has the form

$$F(x) = (x_1^2 + x_2^2)^2 - 2a^2(x_1^2 - x_2^2) = 0, \qquad x = (x_1,\ x_2)^T \qquad (1.66)$$

This curve is shown on Fig. 1.4 by solid line and intersects the x_1 axis at the points $x_1 = \pm a\sqrt{2}$ and $x_1 = 0$. Below we will consider the case $a = 1$. The continuation equations (1.61) take the form

$$\begin{bmatrix} ((x_1^2 + x_2^2)x_1 - a^2x_1) & ((x_1^2 + x_2^2)x_2 + a^2x_2) \\ x_{1,\lambda}^* & x_{2,\lambda}^* \end{bmatrix} \begin{bmatrix} x_{1,\lambda} \\ x_{2,\lambda} \end{bmatrix} = \begin{bmatrix} 0 \\ 1 \end{bmatrix}. \qquad (1.67)$$

The computations were started from the point A, i.e., the initial conditions were taken as

$$x_{(0)} = (\sqrt{2},\ 0)^T. \qquad (1.68)$$

The vector $x,^*_{\lambda(-1)}$ was necessary to start the computations. It was defined as $(0, 1)^T$ since A was a limiting point with respect to variable x_1.

Fig. 1.4 presents the results of the Cauchy problem integration (1.67), (1.68) with the step $\Delta\lambda = 0.2$. The linear equations systems were solved by Gauss method. The dots correspond to the explicit scheme of Euler method (1.62). The crosses designate the results of using the modified Euler method (1.63). The results of using the forth - order Runge - Kutta method and all the implicit schemes are shown by circles.

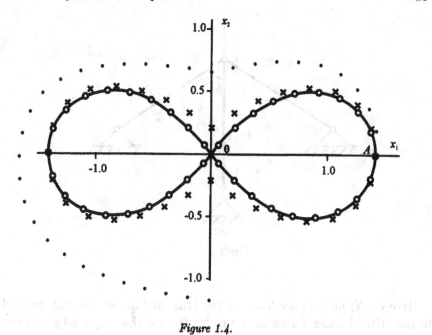

Figure 1.4.

The error that accumulates by using Runge - Kutta method with a step $\Delta\lambda = 0.2$ can be illustrated by the following result: after four passes around the lemniscate we arrived to the point with coordinates $x_1 = 1.4112$, $x_2 = -0.0026$. If there were no error accumulated during the solution, the curve would have come to the point $A(\sqrt{2}, 0)$.

Three - rod truss.

Consider the symmetric deformation of three - rod truss (Fig. 1.5) taking into account the possible buckling of the rods. It is assumed that the length of rods is unit, and the rods are not ideally straight. At initial nondeformed state they are like a half - wave of sinusoid with the amplitudes ε_1 and ε_2 ($|\varepsilon_1| \ll 1$, $|\varepsilon_2| \ll 1$. Indexes 1 and 2 corresponds here and below to numbers of rods. Deformation of such a truss is considered in detail in [48] and is described by the following equations (for the rods with identical cross-sections)

$$N_1 + N_2 - P = 0$$

$$N_1 - 2N_2 + \beta[(W_1^2 - 2W_2^2) - (\varepsilon_1^2 - 2\varepsilon_2^2)] \quad = 0 \qquad (1.69)$$

$$W_1(1 - N_1) - \varepsilon_1 = 0, \qquad W_2(1 - N_2) - \varepsilon_2 = 0$$

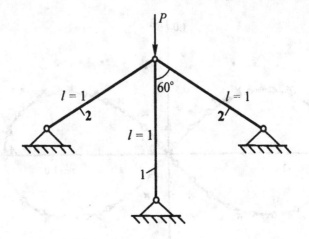

Figure 1.5.

Here P, N_1 and N_2 are loads on the truss and stresses in rods related to the critical Euler forces of the rods. W_i are the amplitudes of total deflections of rods. β is the parameter of rods flexibility.

If $P = 0$ then the nonlinear equations system (1.69) has a trivial solution corresponding to the nondeformed state

$$P = 0, \qquad N_1 = N_2 = 0, \qquad W_1 = \varepsilon_1, \qquad W_2 = \varepsilon_2.$$

If the rods are ideally straight, i.e., when $\varepsilon_1 = \varepsilon_2 = 0$ the system (1.69) has four exact solutions

1) $N_1 = \frac{2}{3}P$, $N_2 = \frac{1}{3}P$, $W_1 = W_2 = 0$;

2) $N_1 = 1$, $N_2 = P - 1$, $W_1^2 = \dfrac{2P - 3}{\beta}$, $W_2 = 0$;

3) $N_1 = P - 1$, $N_2 = 1$, $W_1 = 0$, $W_2^2 = \dfrac{P - 3}{2\beta}$;

4) $N_1 = 1$, $N_2 = 1$, $P = 2$, $W_1^2 - 2W_2^2 = \dfrac{1}{\beta}$.

If $\beta = 100$ the solutions and corresponding deformed configurations of the truss are shown at Fig. 1.6 in the space W_1, W_2, P. From this figure it is seen that the solution set of the system (1.69) varies in a complicated manner in space and has three brunch points B_1, B_2, B_3.

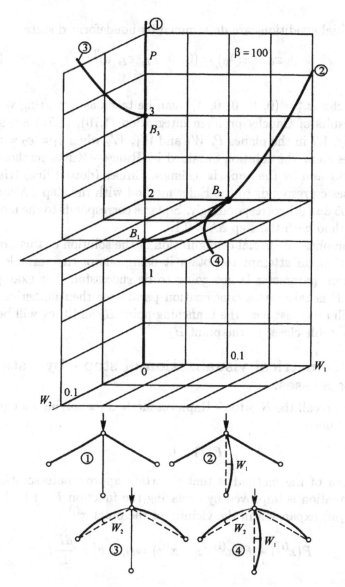

Figure 1.6.

If the vector $x = (N_1,\ N_2,\ W_1,\ W_2,\ P)^T$ is introduced then the continuation equations (1.69) become

$$
\begin{bmatrix}
1 & 1 & 0 & 0 & -1 \\
1 & -2 & 2\beta W_1 & -4\beta W_2 & 0 \\
-W_1 & 0 & 1-N_1 & 0 & 0 \\
0 & -W_2 & 0 & 1-N_2 & 0 \\
x_{1,\lambda}^* & x_{2,\lambda}^* & x_{3,\lambda}^* & x_{4,\lambda}^* & x_{5,\lambda}^*
\end{bmatrix}
x_{,\lambda} =
\begin{bmatrix}
0 \\
0 \\
0 \\
0 \\
1
\end{bmatrix} .
\tag{1.70}
$$

The initial conditions are determined by nondeformed state

$$x_0 = x(\lambda_0) = (0, \ 0, \ \varepsilon_1, \ \varepsilon_2, \ 0)^T. \tag{1.71}$$

The vector $x_{,\lambda}^* = (0, \ 0, \ 0, \ 0, \ 1)^T$ can be taken as a starting vector.

The results of Cauchy problem integration (1.70), (1.71) are shown on the Fig. 1.7 in the planes P, W_1 and W_1, W_2 when $\varepsilon_1 = \varepsilon_2 = 0.001$. Solid lines show the solution obtained by Runge - Kutta method with the step 0.1 and by the implicit schemes. Circles (dotted line) triangles and crosses corresponds to the Euler method with the step $\Delta\lambda$ equal to 0.025, 0.05 and 0.1 correspondingly. Squares corresponds to the modified Euler method with the step $\Delta\lambda = 0.1$.

The complicated special configuration of the solution of this problem suggests that an attempt to obtain it using one of the variables as a continuation parameter is not going to be successful. For example, if the load P is taken as a continuation parameter then numerical difficulties will arise just near the branching point B_1 and they will become insurmountable closer to the point B_2.

5. Geometrical visualization of step - by - step processes.

We first recall the Newton - Raphson method for solving an equation in one unknown

$$F(x) = 0. \tag{1.72}$$

The idea of the method is that a certain approximate solution $x^{(i)}$ of this equation is improved by replacing the function $F(x)$ by its first order Taylor expansion in the vicinity of the point $x^{(i)}$

$$F(x^{(i)}) + F'(x^{(i)})(x - x^{(i)}) = 0, \quad F' = \frac{dF}{dx}. \tag{1.73}$$

The solution of this equation

$$x = x^{(i+1)} = x^{(i)} - (F'(x^{(i)}))^{-1} F(x^{(i)}) \tag{1.74}$$

gives a new approximation $x^{(i+1)}$ of the solution of the equation (1.72). By repeating this operation we construct an iterative process which with a good starting approximation enables us to find the solution of equation (1.72) with the desired accuracy. Fig. 1.8 shows the geometric interpretation of this process. Of course $F' = dF/dx$ must not vanish in the iterative process

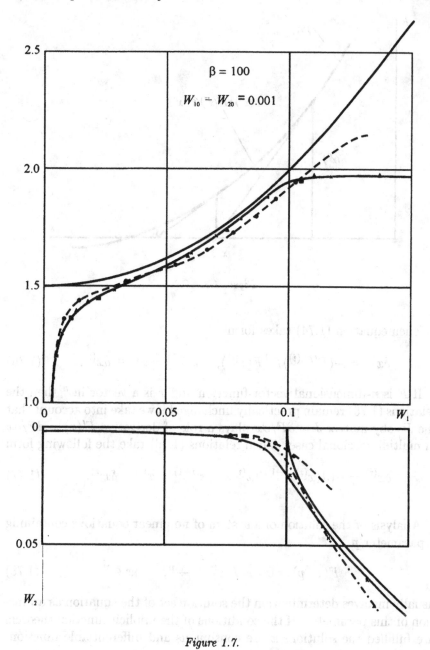

$\beta = 100$

$W_{10} = W_{20} = 0.001$

Figure 1.7.

Sometimes it is convenient to write equation (1.73) as an equation for the correction $\delta x^{(i)} = x^{(i+1)} - x^{(i)}$ that improves the solution by i - th iteration

$$F(x^{(i)}) + F'(x^{(i)})\delta x^{(i)} = 0. \qquad (1.75)$$

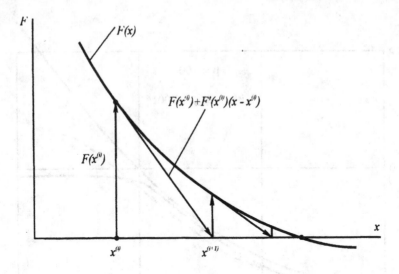

Figure 1.8.

Then equation (1.74) takes form

$$\delta x^{(i)} = -(F'(x^{(i)}))^{-1}F(x^{(i)}), \quad x^{(i+1)} = x^{(i)} + \delta x^{(i)}. \quad (1.76)$$

If F is n-dimensional vector-function and x is a vector in ^{n} then the relations (1.76) remain practically unchanged if we take into account that the Jacoby matrix $J = \partial F/\partial x$ plays a role of derivative $F'(x) = dF/dx$ in multidimensional case. Then relations (1.76) take the following form

$$\delta x^{(i)} = -(J(x^{(i)}))^{-1}F(x^{(i)}), \quad x^{(i+1)} = x^{(i)} + \delta x^{(i)}. \quad (1.77)$$

Analysis of the solution of a system of nonlinear equations containing a parameter p

$$F(x,\, p) = 0 \qquad F : ^{n+1} \to\, ^n, \qquad x \in\, ^n \qquad (1.78)$$

usually involves determination the solution set of the equation as a function of this parameter. If the conditions of the implicit function theorem are fulfilled the solutions x are continuous and differentiable functions of p, i.e., $x = x(p)$.

The simplest form of representing this relation would be to find solutions $x_{(k)} = x(p_k)$ for certain set of parameter values $p : p_0 < p_1 < \ldots < p_k < \ldots < p_N$. Lahaye process (1.5) demonstrates how the solutions construction process can be designed geometrically within the framework of continuation idea. In fact, he suggested to use the solution for the previous value of the parameter p as a starting approximation for the current value of the parameter. This process can be written for $p = p_k$ as

$$x_{(k)}^{(0)} = x_{(k-1)},$$

$$\delta x_{(k)}^{(i)} = -(J(x_{(k)}^{(i)}))^{-1} F(x_{(k)}^{(i)}, p_k), \quad x_{(k)}^{(i+1)} = x_{(k)}^{(i)} + \delta x_{(k)}^{(i)}, \quad (1.79)$$

$$i = 0, 1, 2, \ldots$$

The process continues as long as the norm of the correction vector $\|\delta x_{(k)}^{(i)}\|$ exceeds a given accuracy $\varepsilon > 0$.

Fig. 1.9 shows the geometry of this process of going from p_{k-1} to p_k for an equation in one unknown. The desired solution set of the equation $F(x, p) = 0$ is a curve K along which the surface $F = F(x, p)$ in space $^3 : \{x, p, F\}$ intersects the plane $^2 : \{x, p\}$. The Newton - Raphson iterative process takes place in the plane $p = p_k$ with the starting approximation $x_{(k)}^{(0)} = x_{(k-1)}$.

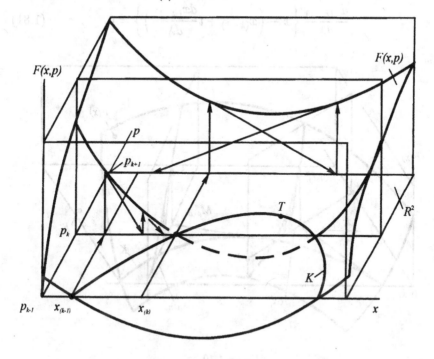

Figure 1.9.

The same figure shows clearly that the difficulties arising in the vicinity of the limiting point T are connected with passage from p_k to p_{k+1}

which brings the process (1.79) out of the domain where the solution exists. Apparently, these difficulties are due to the fact that the solution is sought in the plane $p = p_{k+1}$ which has no intersection with the solution curve K. These difficulties can be avoided if for each k we design the Newton - Raphson iterative process of seeking $x_{(k)}$ in a plane $M_{(k)}$ which is orthogonal to the curve K for $x = x_{(k)}$. However, the plane $M_{(k)}$ is not known until $x_{(k)}$ is found. Nevertheless, the solution can be sought in a plane $M_{(k)}^*$ close to $M_{(k)}$.

We consider now one of the ways of specifying $M_{(k)}^*$. We introduce a vector $x = (x_1, \ x_2 = p)^T$ and consider the equation

$$F(x) = 0. \tag{1.80}$$

Let t be a step size with which we try to move along the curve K. A plane close to the plane $M_{(k)}$ is then a plane $M_{(k)}^*$ passing through a point $(x_{(k-1)} + t \, dx_{(k-1)}/d\lambda) \in {}^{n+1}$ so that it is orthogonal to the unit vector $dx_{(k-1)}/d\lambda$ tangential to the curve K at the previous point $x_{(k-1)}$ (Fig. 1.10). The plane $M_{(k)}^*$ is therefore defined by the equation

$$\frac{dx_{(k-1)}}{d\lambda} \left\{ x - \left(x_{(k-1)} + t \frac{dx_{(k-1)}}{d\lambda} \right) \right\} = 0. \tag{1.81}$$

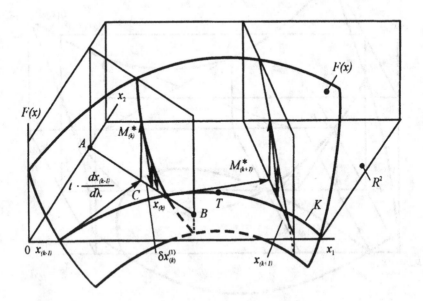

Figure 1.10.

The vector in the parentheses corresponds to the point C at Fig. 1.10. And the vector in braces connects the point C with an arbitrary point of the plane $^2 : \{x_1, x_2\}$. If it is orthogonal to the vector $dx_{(k-1)}/d\lambda$ as the equation (1.81) requires, it will belong to the straight line AB.

Thus the determination of the solution $x_{(k)}$ is reduced to finding the solution of equation (1.80) in the plane $M^*_{(k)}$, i.e. , to the simultaneous solution of equation (1.80) and (1.81). If we simplify the second equation taking into account that the vector $dx_{(k-1)}/d\lambda$ is a unit then this system takes the form

$$F(x) = 0,$$

$$\frac{dx_{(k-1)}}{d\lambda}(x - x_{(k-1)}) = t. \tag{1.82}$$

The iterative process of the Newton - Raphson method for this system is as follows

$$x_{(k)}^{(0)} = x_{(k-1)} + t\frac{dx_{(k-1)}}{d\lambda},$$

$$\delta x_{(k)}^{(i)} = - \begin{bmatrix} J(x_{(k)}^{(i)}) \\ \frac{dx_{(k-1)}}{d\lambda} \end{bmatrix}^{-1} \begin{bmatrix} F(x_{(k)}^{(i)}) \\ \frac{dx_{(k-1)}}{d\lambda}(x_{(k)}^{(i)} - x_{(k-1)}) - t \end{bmatrix}, \tag{1.83}$$

$$x_{(k)}^{(i+1)} = x_{(k)}^{(i)} + \delta x_{(k)}^{(i)}, \quad i = 0, 1, 2, \ldots$$

Note that the vector $x_{(k)}^{(i)} - x_{(k-1)}$ begins at the point $x_{(k-1)}$ and it ends on the straight line AB. Then its projection on the vector $dx_{(k-1)}/d\lambda$, i.e. the scalar product of these vectors, is just equal to t. Therefore the last component of the vector in right - hand side of equation (1.81) is equal to zero and the equation itself takes a simpler form

$$\delta x_{(k)}^{(i)} = - \begin{bmatrix} J(x_{(k)}^{(i)}) \\ \frac{dx_{(k-1)}}{d\lambda} \end{bmatrix}^{-1} \begin{bmatrix} F(x_{(k)}^{(i)}) \\ 0 \end{bmatrix}. \tag{1.84}$$

Taking this into account, the second equation (1.82) can be written as

$$\frac{dx_{(k-1)}}{d\lambda}\delta x_{(k)}^{(i)} = 0. \tag{1.85}$$

Geometrically this equation implies that the correction vector $\delta x_{(k)}^{(i)}$ is orthogonal to the unit vector $dx_{(k-1)}/d\lambda$ (see Fig. 1.10). This interpretation of the additional equation (1.82) suggests iterative processes which are expected to be more efficient. Thus, if we introduce the vector

$$\xi_{(k)}^{(i)} = x_{(k)}^{(i)} - x_{(k-1)},$$

then by analogy with (1.85) the additional condition in (1.82) can be formulated as

$$\xi_{(k)}^{(i)} \delta x_{(k)}^{(i)} = 0.$$

The geometry of an iterative process with such a condition is shown in the Fig. 1.11. In this process the position of the plane M_k^*, in which the solution is sought, is corrected at each iteration.

If in addition the vector $\xi_{(k)}^{(i)}$ is normalized at each iteration so that it has the length t we obtain an iterative process illustrated in the plane $^2 : \{x_1, \ x_2\}$ of the Fig. 1.12. Its algorithm is of the form

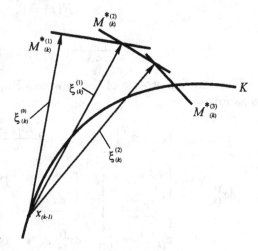

Figure 1.11.

$$\xi_{(k)}^{(0)} = t\frac{dx_{(k-1)}}{d\lambda}, \quad x_{(k)}^{(0)} = x_{(k-1)} + \xi_{(k)}^{(0)},$$

$$\delta x_{(k)}^{(i)} = - \begin{bmatrix} J(x_{(k)}^{(i)}) \\ \xi_{(k)}^{(i)} \end{bmatrix}^{-1} \begin{bmatrix} F(x_{(k)}^{(i)}) \\ 0 \end{bmatrix},$$

$$\xi_{(k)}^{(i+1)} = t\frac{\xi_{(k)}^{(i)} + \delta x_{(k)}^{(i)}}{\|\xi_{(k)}^{(i)} + \delta x_{(k)}^{(i)}\|}, \quad x_{(k)}^{(i+1)} = x_{(k-1)} + \xi_{(k)}^{(i)}, \quad i = 0, 1, 2, \ldots$$

$$(1.86)$$

Note that all the foregoing algorithms admit a usual for the Newton - Raphson method modification. This modification replaces the matrix $J(x_{(k)}^{(i)})$, which varies from iteration to iteration, with the first order approximation matrix $J(x_{(k)}^{(0)})$. Pay attention to the fact that all the iterative processes described above are essentially processes for solving equation (1.72) simultaneously with some additional condition. Thus, the simplest additional condition $x_2 = p_k$ is used in the Lahaye process (1.79), the condition (1.81) is used in the process (1.83), and so

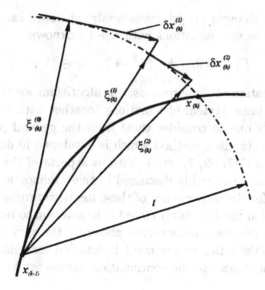

Figure 1.12.

on. All these conditions determine in ^{2} a line of intersection of the plane $M_{(k)}^*$ which may vary from iteration to iteration with the plane $^2 : \{x_1, x_2\}$. Generalizing this approach we can formulate an additional condition defining a line in $^2 : \{x_1, x_2\}$ and seek the solution of equation (1.80) as the point of intersection of the solution set curve K with this line. Thus if a circle of radius t with the point $x_{(k-1)}$ as its center is chosen as such a line we arrive to simultaneous solution of the following equations

$$F(x) = 0, \quad (x - x_{(k-1)})(x - x_{(k-1)}) - t^2 = 0. \qquad (1.87)$$

The algorithm of the Newton - Raphson method for this system of equations is of the form

$$\xi_{(k)}^{(0)} = t\frac{dx_{(k-1)}}{d\lambda}, \quad x_{(k)}^{(0)} = x_{(k-1)} + \xi_{(k)}^{(0)},$$

$$\delta x_{(k)}^{(i)} = - \begin{bmatrix} J(x_{(k)}^{(i)}) \\ \xi_{(k)}^{(i)} \end{bmatrix}^{-1} \begin{bmatrix} F(x_{(k)}^{(i)}) \\ \frac{1}{2}(t^2 - \xi_{(k)}^{(i)}\xi_{(k)}^{(i)}) \end{bmatrix}, \qquad (1.88)$$

$$x_{(k)}^{(i+1)} = x_{(k)}^{(i)} + \delta x_{(k)}^{(i)}, \quad \xi_{(k)}^{(i+1)} = x_{(k)}^{(i+1)} - x_{(k-1)}, \quad i = 0, 1, 2, \ldots$$

Without any changes all processes analyzed above can be used for solving the system of n equations with $n+1$ unknowns

$$F(x) = 0, \quad F:^{n+1} \to ^n, \quad x \in ^{n+1}.$$

The interpretation of these processes as algorithms for the simultaneous solution of basic system of equations together with the additional equation enables one to consider them from the general point of view of the Newton - Raphson method which is developed in detail in many monographs (see [111, 62, 7], etc.) Various aspects of the convergence of the method are thoroughly discussed in these monographs. Here we only note that for the convergence of these iterative processes the starting approximation must be fairly close to the solution to be determined. In the iterative processes constructed above by the very nature of the continuation method this requirement is satisfied for sufficiently small step sizes t with respect to the continuation parameter λ.

6. The solution continuation in vicinity of essential singularity points

The development of a continuation algorithm in the vicinity of essential singularities is not a goal of this book. The methods of the solution of this problem are specific to the solution behavior near these points. At present this problem is solved only for certain functions. The analysis of the problem from the point of view of the solution continuation is given in monograph [48] and in the chapter 8 of the present book. Here we consider only the simplest case of analysis of the plane curve singularities. This allows us to examine briefly the arising problems.

The plane curve corresponds to the solution set of one equation with two unknowns

$$F(x_1, x_2) = 0. \tag{1.89}$$

Here the function $F(x_1, x_2)$ is assumed to be continuous and sufficiently smooth with respect to two variables in the vicinity of the considered point $x^0 = (x_1^0, x_2^0) \in ^2$ which satisfies the equation (1.89), i.e.,

$$F(x_1^0, x_2^0) = 0. \tag{1.90}$$

For the sake of simplicity we denote derivatives as $\partial F/\partial x_i = F_{,i}$. Then the Jacoby matrix J of the function $F(x_1, x_2)$ is of the form

$$J = (F_{,1}, F_{,2}) = 0. \tag{1.91}$$

The solution can be continued if at least one of the derivatives $F_{,i}$ is not equal to zero. Thus, if the x_2 is taken as a parameter, the points where $F_{,1} \neq 0$ are regular and the points where $F_{,1} = 0$, $F_{,2} \neq 0$ are limiting. But $rank(J) = 1$ for all these points.

If in the point x^0

$$F_{,1}(x_1^0, x_2^0) = 0, \quad F_{,2}(x_1^0, x_2^0) = 0, \qquad (1.92)$$

then $rank(J) = 0$ in this point, i.e., the matrix J is degenerate, and the point is singular. By virtue of equations (1.90), (1.92) Taylor series expansion of the function in vicinity of this point is represented as

$$F(x_1, x_2) = \frac{1}{2!}(F_{,11}^0 \Delta x_1^2 + 2F_{,12}^0 \Delta x_1 \Delta x_2 + F_{,22}^0 \Delta x_2^2) + O(\rho^3) \quad (1.93)$$

Here $\Delta x_i = x_i - x_i^0$, $\rho = \sqrt{\Delta x_1^2 + \Delta x_2^2}$, $F_{,ij}^0 = F_{,ij}(x_1^0, x_2^0)$, $i, j = 1, 2$.

In accordance to the representation (1.93) the sum of the second - degree terms in the expansion must vanish in the vicinity of the point x^0 on the tangent to the solution set curve K, i.e.,

$$F_{,11}^0 \Delta x_1^2 + 2F_{,12}^0 \Delta x_1 \Delta x_2 + F_{,22}^0 \Delta x_2^2 = 0. \qquad (1.94)$$

If, for example, $F_{,22}^0 \neq 0$ the tangent to the curve K can be specified by

$$\Delta x_2 = t \Delta x_1. \qquad (1.95)$$

Then from (1.94) it follows that

$$F_{,11}^0 + 2F_{,12}^0 t + F_{,22}^0 t^2 = 0. \qquad (1.96)$$

The tangent exists if this equation has real roots. The number of roots is determined by the discriminant

$$D = F_{,11}^0 F_{,22}^0 - (F_{,12}^0)^2. \qquad (1.97)$$

If $D < 0$ the equation has two distinct real roots t_1, t_2 which in accordance with (1.95) give two distinct values of the derivative at the point x^0

$$\frac{dx_2}{dx_1} = t_1, \qquad \frac{dx_2}{dx_1} = t_2.$$

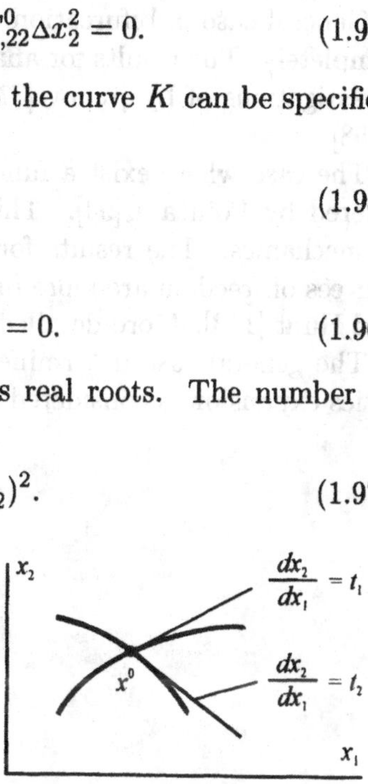

Figure 1.13.

Thus, through the point x_0 pass two curve the tangents to which are determined by expressions (1.95) when $t = t_1$ and $t = t_2$. Such point are the branching point. See Fig. 1.13.

If $D > 0$ equation (1.96) has no real roots. This means that the tangent to the curve at this point does not exist and the curve itself

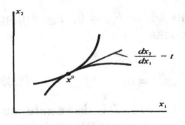
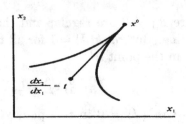

Figure 1.14.   *Figure 1.15.*

does not exist too at the point. So this point is an isolated point and it is not possible to arrive to it continuing the solution.

Finally for $D = 0$ equation (1.96) has two real roots t. In this case the true behavior of the curve at the point x^0 can only be established by analyzing the third and the higher order terms in the Taylor series expansion. It is possible that here the singular point of two tangential curves (Fig. 1.14) or a cusp (Fig. 1.15).

General case in bifurcation of curves in $^{n+1}$ has not yet been studied completely. The results for analytic functions F_i which date back to the investigations of Lyapunov [77] and Shmidt [103] are given in [117, 62, 3, 58].

The case where exist a function F such that $F_i = \partial F / \partial x_i$ was considered by Poincare [94]. This is of particular interest to application in mechanics. The results for elastic systems with a finite number of degrees of freedom are more fully presented in monograph of Thompson and Hunt [113]. More detailed literature reference are given in [48].

The general case determined by the second - order terms at Taylor series expansion is considered in chapter 8 of the present book.

Chapter 2

THE CAUCHY PROBLEM FOR ORDINARY DIFFERENTIAL EQUATIONS

We consider in this chapter the Cauchy problem for a system of ordinary differential equations (ODE). Under certain conditions the solution of this problem is a smooth integral curve in the space of unknowns and parameter, i.e., a one - parametric set similar to those considered at the previous chapter. This allows us to consider the Cauchy problem within the framework of the method of parametric continuation. Such a view leads to the formulation of the best continuation parameter problem, which is solved below.

1. The Cauchy problem as a problem of solution continuation with respect to a parameter

Consider the Cauchy problem for normal system of ODE

$$\frac{dy_i}{dt} = f_i(y_1, y_2, \ldots, y_n, t), \qquad y_i(t_0) = y_{i0}, \qquad i = \overline{1, n}. \qquad (2.1)$$

Assume that the conditions of existence and the uniqueness theorem are fulfilled for this problem.

Let an integral the problem to be given by the relations

$$F_j(y_1, \ldots, y_n, t) \quad = \quad 0, \qquad j = \overline{1, n},$$

$$F_j(y_{10}, \ldots, y_{n0}, t_0) \quad = \quad 0, \qquad (2.2)$$

that define the integral curve of the problem (2.1) in $(n+1)$ - dimensional Euclidean space $^{n+1} : \{y_1, \ldots, y_n, t\}$.

The process of finding this curve can be considered as a problem of constructing of the solutions set of the system of equations (2.2) containing the parameter - argument t for various values of t. We will use parametric continuation method to solve this system. Then the problem (2.1) can be considered as a Cauchy problem for the continuation equations of the system (2.2) solution with respect to the parameter t when the system is reduced to the normal form. Hence, similar to the chapter 1, we can look for the best continuation parameter. We will call it the best argument.

As in chapter 1 we choose the best argument - parameter locally, i.e.,in a small vicinity of each point of the solution set curve — integral curve of the problem (2.1).

To solve the problem we assume that y_i and t are such functions of some argument μ that at each point of the integral curve of the problem

$$d\mu = \alpha_i dy_i + \alpha_{n+1} dt, \qquad i = \overline{1, n}. \tag{2.3}$$

Here $\alpha_k (k = \overline{1, n+1})$ are the components of a unit vector $\alpha = (\alpha_1, \ldots, \alpha_{n+1})^T$ determining the direction along which the argument μ is measured. Note that the summation with respect to repeated subscript i is assumed in the expression (2.3).

Right - hand side of the equality (2.3) can be regarded as the scalar product of the vector α and the vector - function differential $(dy_1, \ldots, dy_n, dt)^T$. Assigning various values to the components of the vector α it is possible to consider all possible continuation parameters of the problem (2.2), i.e., all arguments of the problem (2.1).

Since the particular form of equation (2.2) is unknown the change to the argument μ can be implemented immediately for the problem (2.1). Dividing the equality (2.3) by $d\mu$ after that, we obtain

$$y_{i,\mu} - f_i t_{,\mu} = 0,$$
$$\alpha_i y_{i,\mu} + \alpha_{n+1} t_{,\mu} = 1 \qquad i = \overline{1, n}. \tag{2.4}$$

If the vector $y = (y_1, \ldots, y_n, t)^T \in {}^{n+1}$ is introduced, the system (2.4) can be written in the matrix form

$$\begin{bmatrix} A \\ \alpha \end{bmatrix} y_{,\mu} = \begin{bmatrix} 0 \\ 1 \end{bmatrix}. \tag{2.5}$$

Here the matrix A of size $n \times (n+1)$ has the structure

$$A = [Ef],$$

where E is the unit matrix of n - th order and $f = (f_1, \ldots, f_n)^T$ is a vector in ^{n}.

The structure of the system (2.5) is exactly the same as that of the system (1.37). Therefore, in according with the Theorem 1, the transition to the normal form of the system (2.5) is the best conditioned if and only if $\alpha = y_{,\lambda}$, i.e., if the arc length of the curve of the system (2.2) solution is chosen as the parameter μ. This curve is the integral curve of the problem (2.1). Thus, the system (2.5) can be written in the form

$$y_{i,\lambda} y_{i,\lambda} + t_{,\lambda}^2 = 1$$
$$y_{i,\lambda} - f_i t_{,\lambda} = 0,$$

(2.6)

which can be solved analytically with respect to derivatives. Since the argument does not appear explicitly in the equations we take the initial point of the Cauchy problem (2.1) as the initial point of λ. Then we arrive to the following form of the Cauchy problem [106]

$$\frac{dy_i}{d\lambda} = \frac{f_i}{\sqrt{1 + f_j f_j}} \quad y_i(0) = y_{i0}$$
$$\frac{dt}{d\lambda} = \frac{1}{\sqrt{1 + f_j f_j}} \quad t(0) = t_0$$

(2.7)

$$i, j = \overline{1, n}.$$

Below, the argument λ, which provides the best conditioning to the system of equation (2.5), will be named as the best argument.

Thus the main result of this chapter has been proved

Theorem 2. For the Cauchy problem for the normal system of ODE (2.1) to be transformed to the best argument it is necessary and sufficient that the arc length of the solution curve to be chosen as this argument. In this case the problem (2.1) is transformed to the problem (2.7).

The new formulation of the Cauchy problem (2.7) has a number advantages in comparison with the Cauchy problem (2.1). First, the right - hand side of each equation (2.7) does not exceed unit. Moreover, the squared norm of the system right - hand sides is always equal to unit. This removes many of the problems connected with unlimited growth of the right - hand sides of the system (2.1), and allows to integrate differential equations which have the

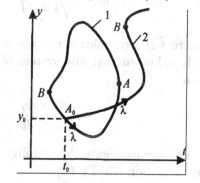

Figure 2.1.

limiting points at integral curves (points A and B, at Fig. 2.1) where the derivatives become infinite. It becomes possible to solve problem with closed integral curves (curve 1, at Fig. 2.1). Also, it will be shown later that the suggested transformation reduces the difficulties that are typical for stiff systems.

Note that the advantages of the λ - transformation are not only numerical. It can be used successfully in the qualitative theory of differential equations transforming the space which contains the unlimited functions in the system right-hand sides into the space where they are limited. Such transition was made in [92] in investigation of the autonomous systems of ODE. The normalization proposed in [92] changes the integral curve of the problem without changing its phase space. The λ - transformation do not have this disadvantage. Finally, the lemmas 2 and 3 state that the choice of the best parameter as the argument minimizes the quadratic error resulting from perturbations of matrix elements and right - hand sides of the system (2.5). It is particularly important for numerical integration of the normal system of ODE (2.7) obtained from the implicit system (2.6) because the numerical errors will have minimal effect on the solution.

2.   Certain properties of - transformation

Before considering properties of the obtained transformation note that in the case when function in the right - hand side of the system (2.1) is complex - valued and the desired solution is real the transformed problem (2.7) has the form

$$\frac{dy_i}{d\lambda} = \frac{f_i}{\sqrt{1 + f\bar{f}}}, \quad y_i(0) = y_{i0},$$

$$\frac{dt}{d\lambda} = \frac{1}{\sqrt{1 + f\bar{f}}}, \quad t(0) = t_0, \quad i = \overline{1, n},$$

where $\bar{f}$ is a complex conjugate to the function f.

It is known that any system of ODE can be written in autonomous form

$$\frac{dy}{dt} = f(y). \qquad (2.8)$$

Let us investigate the stability of Euler method for the well - known test equation [23, 72, 64]

$$\frac{dx}{dt} = ax, \qquad (2.9)$$

which is obtained if the function of the right - hand side of the equation (2.8) is represented in vicinity of the point $y = y_m$ by means of Taylor formula as $f(y) = f(y_m) + a(y - y_m)$. The value $a = df(y_m)/dy$ is, in general, a complex constant and functions x and y are related as follows

$$x = y - y_m + \frac{f_m}{a}. \qquad (2.10)$$

Here and below the notation $f_m = f(y_m)$ is used.

Note the following: since we will investigate only real solutions of the equations (2.8), (2.9), i.e., the real values of the functions x and y, it follows from equality (2.10) that the function f_m/a must be real too.

The Euler's explicit scheme for the equation (2.9) is of the form

$$x_{m+1} = x_m + h_t a x_m = (1 + ah_t)x_m. \qquad (2.11)$$

Here h_t is the integration step with respect to t. It is known (see [81] as an example) that this scheme is absolutely stable if

$$|1 + ah_t| < 1 \quad (1 + h_t\alpha)^2 + (h_t\beta)^2 < 1, \qquad (2.12)$$

where $\alpha = Re\, a$, $\beta = Im\, a$ are real and imaginary parts of a.

Thus, the range of absolute stability of the scheme (2.11) on the plane $h_t\alpha$, $h_t\beta$ is the interior of a unit circle with its center at the point $(-1, 0)$. This is demonstrated at the Fig. 2.2,a.

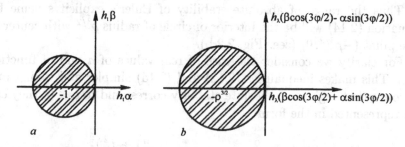

Figure 2.2.

Using the λ - transformation for the equation (2.8) we obtain

$$\frac{dy}{d\lambda} = \frac{f(y)}{\sqrt{1 + f(y)\overline{f(y)}}}, \qquad \frac{dt}{d\lambda} = \frac{1}{\sqrt{1 + f(y)\overline{f(y)}}}. \qquad (2.13)$$

In this system of two equations the solution of the second equation is determined by the solution of the first one. That is why we consider only the first equation in (2.13). It is easily to establish the equality $a\overline{f_m} = \overline{a}f_m$ which follows from the condition that the solution of equations (2.8), (2.9) are real.

If the right - hand side of the first equation of the system (2.13) is linearized by the Taylor's formula in the vicinity of the value $y = y_m$ then we obtain

$$\frac{dz}{d\lambda} = \frac{a}{(1 + f_m \overline{f_m})^{\frac{3}{2}}} z. \tag{2.14}$$

Here $z = y - y_m + f_m(1 + f_m \overline{f_m})/a = x + f_m^2 \overline{f_m}/a$ is a new function.

The range of the absolute stability of the explicit Euler's scheme for this equation is defined by the inequality

$$\left| 1 + \frac{a h_\lambda}{(1 + f_m \overline{f_m})^{\frac{3}{2}}} \right| < 1. \tag{2.15}$$

Here h_λ is the integration step with respect to λ. Introduce the notation

$$\rho = 1 + f_m \overline{f_m} \geq 1.$$

Then inequality (2.15) can be reduced to the form

$$\left(\rho^{\frac{3}{2}} + \alpha h_\lambda \right)^2 + (\beta h_\lambda)^2 < \left(\rho^{\frac{3}{2}} \right)^2.$$

Thus the range of absolute stability of Euler's explicit scheme for equation (2.14) will be the interior of circle of radius $\rho^{3/2}$ with center at the point $(-\rho^{3/2}, 0)$ (see. Fig. 2.2,b).

For clarity we consider the case of real values of a and of function f_m. This makes inequalities (2.12) and (2.15) simpler and taking into account that regions of absolute stability correspond to $a < 0$, they can be represented in the form

$$0 < h_t < \frac{2}{|a|}, \qquad 0 < h_\lambda < \frac{2(1 + f_m^2)^{\frac{3}{2}}}{|a|}. \tag{2.16}$$

It is clear now that λ - transformation extends the range of absolute stability of Euler explicit scheme.

The usage of λ - transformation is justified if it reduces the number of steps leading from the initial point $A_0(t_0, y_0)$ to the final point B (Fig. 2.3) comparably to that for the original parameter t. In other words λ - transformation is effective if at each point of the curve $y(t)$

$$H_\lambda \cos \theta \geq H_t, \tag{2.17}$$

where θ is the angle between the tangent to the curve $y(t)$ and t - axis (Fig. 2.3); H_t, H_λ are the smallest integration steps with respect to the

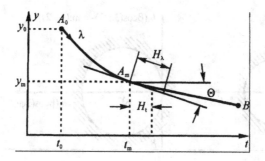

Figure 2.3.

variables t and λ, respectively, for which the iterative process described by Euler's explicit formula ceases to converge. If a is a real value then according to (2.16) we have at the point A_m

$$H_t = \frac{2}{|a|}, \qquad H_\lambda = \frac{2(1 + f_m^2)^{\frac{3}{2}}}{|a|}.$$

Since

$$\cos\theta = \left[1 + \left(\frac{dy}{dt}\right)^2\right]^{-\frac{1}{2}} = (1 + f_m^2)^{-\frac{1}{2}},$$

we have

$$\frac{H_\lambda \cos\theta}{H_t} = 1 + f_m^2 \geq 1.$$

This proves inequality (2.17) and therefore the effectiveness of the λ - transformation.

We will now study the effect of the λ - transformation on the stability of the implicit scheme of Euler's method. For equation (2.9) this scheme has the form

$$x_{m+1} = x_m + h_t a x_{m+1}, \qquad x_{m+1} = x_m(1 - h_t a)^{-1}.$$

The solution will be absolutely stable provided that

$$|(1 - h_t a)^{-1}| < 1 \quad \text{or} \quad |1 - h_t a| > 1.$$

Thus the range of stability in the plane $h_t\alpha$, $h_t\beta$ will be the exterior of the unit circle with its center at the point $(1, 0)$ (Fig. 2.4,a).

The range of stability of Euler's implicit scheme is defined by inequality

$$\left|1 - \frac{ah_\lambda}{(1 + f_m\overline{f_m})^{\frac{3}{2}}}\right| > 1,$$

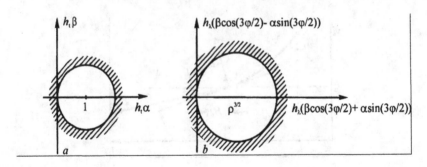

Figure 2.4.

which can be transformed to

$$\left(\rho^{\frac{3}{2}} - \alpha h_\lambda\right)^2 + (\beta h_\lambda)^2 > \left(\rho^{\frac{3}{2}}\right)^2.$$

It defines the exterior of a circle of radius $\rho^{3/2}$ with center at the point $(\rho^{3/2}, 0)$, (Fig. 2.4,b).

If a and function f_m are real, the Euler's implicit scheme can be unstable only when $a > 0$. The range of instability for equations (2.9) and (2.14) is defined by the inequalities

$$0 \leq h_t \leq \frac{2}{a}, \qquad 0 \leq h_\lambda \leq \frac{2(1 + f_m^2)^{\frac{3}{2}}}{a}.$$

Thus the λ - transformation extends the range of stability of the explicit scheme of Euler's method and the range of instability of the implicit scheme.

Note that even though the range of stability of the implicit scheme is reduced the stability still remains A - stable (in terminology of Dahlquist [23]) and stiff stable (in the sense of Geer [42]).

We will now investigate how the λ - transformation changes the spectral characteristics of the system of two differential equations

$$\frac{dy_1}{dt} = f_1(y_1), \qquad \frac{dy_2}{dt} = f_2(y_2). \tag{2.18}$$

We linearize the functions on the right - hand sides of the system (2.18) in the vicinity of values $y_1 = y_{1m}$, $y_2 = y_{2m}$

$$f_1(y_1) = f_{1m} + a_1(y_1 - y_{1m}), \quad f_2(y_2) = f_{2m} + a_2(y_2 - y_{2m}),$$

where $f_{1m} = f_1(y_{1m})$, $f_{2m} = f_2(y_{2m})$, $a_1 = df_1(y_{1m})/dy_1$, $a_2 = df_2(y_{2m})/dy_2$.

If new arguments x_1, x_2 are introduced by formulas

$$x_1 = y_1 - y_{1m} + f_{1m}/a_1, \quad x_2 = y_2 - y_{2m} + f_{2m}/a_2,$$

we obtain the following system of two equations

$$\frac{dx_1}{dt} = a_1 x_1, \qquad \frac{dx_2}{dt} = a_2 x_2. \qquad (2.19)$$

It is clear that, in general, the complex - values a_1, a_2 are the eigenvalues of this system. Applying the λ - transformation to equations (2.18) we obtain the following system

$$\frac{dy_1}{d\lambda} = \frac{f_1}{\sqrt{Q}}, \quad \frac{dy_2}{d\lambda} = \frac{f_2}{\sqrt{Q}}, \quad \frac{dt}{d\lambda} = \frac{1}{\sqrt{Q}},$$

$$Q = 1 + f_1\overline{f_1} + f_2\overline{f_2}. \qquad (2.20)$$

The first two equations are the most important here because the right - hand side of the third equation follows from the first two. We consider the behavior of the system in a small vicinity of the point (y_{1m}, y_{2m}, t_m) on the integral curve corresponding to the value $\lambda = \lambda_m$. Then keeping only the linear part of the Taylor expansions of the right - hand sides we obtain the following relations which characterize the behavior of the system (2.20) in a small vicinity of the point (y_{1m}, y_{2m}, t_m)

$$\frac{dy_1}{d\lambda} = \frac{f_{1m}}{Q^{\frac{1}{2}}} + \frac{a_1(1 + f_{2m}\overline{f_{2m}})}{Q^{\frac{3}{2}}}(y_1 - y_{1m}) - \frac{a_2 f_{1m}\overline{f_{2m}}}{Q^{\frac{3}{2}}}(y_2 - y_{2m}),$$

$$\frac{dy_2}{d\lambda} = \frac{f_{2m}}{Q^{\frac{1}{2}}} - \frac{a_1 f_{2m}\overline{f_{1m}}}{Q^{\frac{3}{2}}}(y_1 - y_{1m}) + \frac{a_2(1 + f_{1m}\overline{f_{1m}})}{Q^{\frac{3}{2}}}(y_2 - y_{2m}),$$

Deriving these equations we have used the following equalities $f_{1m}\overline{a_1} = a_1\overline{f_{1m}}$, $f_{2m}\overline{a_2} = a_2\overline{f_{2m}}$, which follow from the fact that the functions y_1, y_2, x_1, x_2 are real.

The latter system of nonhomogeneous linear ODE can be transformed to the homogeneous system if the new functions

$$z_i = y_i + \delta_i, \qquad i = 1, 2,$$

are introduced. Here members δ_i are real roots of the following system of linear algebraic equations

$$a_{11}\delta_1 + a_{12}\delta_2 = b_1,$$
$$a_{21}\delta_1 + a_{22}\delta_2 = b_2,$$

The coefficients of the system are calculated as

$$a_{11} = a_1(1 + f_{2m}\overline{f_{2m}}), \qquad a_{12} = -a_2 f_{1m}\overline{f_{2m}},$$

$$a_{21} = -a_1 f_{2m}\overline{f_{1m}}, \qquad a_{22} = a_2(1 + f_{1m}\overline{f_{1m}}), \qquad (2.21)$$

$$b_1 = f_{1m}Q - a_{11}y_{1m} - a_{12}y_{2m}, \qquad b_2 = f_{2m}Q - a_{21}f_{1m} - a_{22}y_{2m}.$$

Then in a vicinity of the point (y_{1m}, y_{2m}, t_m) the first two equations of the ODE system (2.20) can be represented in the form

$$\frac{dz_1}{d\lambda} = q(a_{11}z_1 + a_{12}z_2), \qquad \frac{dz_2}{d\lambda} = q(a_{21}z_1 + a_{22}z_2), \qquad (2.22)$$

where $q = Q^{-\frac{3}{2}}$.

The spectrum of eigenvalues of the system (2.22) consist of the roots of the characteristic equation

$$\begin{vmatrix} a_{11} - \dfrac{r}{q} & a_{12} \\ a_{21} & a_{22} - \dfrac{r}{q} \end{vmatrix} = 0,$$

which are calculated as

$$r_{1,2} = \frac{q}{2}\left(a_{11} + a_{22} \pm \sqrt{(a_{11} - a_{22})^2 + 4a_{12}a_{21}}\right).$$

After introduction of the parameter $\varepsilon = a_2/a_1$ we obtain

$$r_{1,2} = \frac{q}{2}\left(a_{11} + a_{22} \pm (a_{11} - a_{22})\sqrt{1 + \frac{4a_1 a_2 f_{1m}\overline{f_{1m}} f_{2m}\overline{f_{2m}}}{(a_{11} - a_{22})^2}}\right) =$$

$$= \frac{q}{2}\left(a_{11} + a_{22} \pm (a_{11} - a_{22})\cdot\right.$$

$$\left.\sqrt{1 + \varepsilon\frac{4 f_{1m}\overline{f_{1m}} f_{2m}\overline{f_{2m}}}{(1 + f_{2m}\overline{f_{2m}})^2\left(1 - \varepsilon\dfrac{1 + f_{1m}\overline{f_{1m}}}{1 + f_{2m}\overline{f_{2m}}}\right)^2}}\right).$$

The use the following notations

$$\varepsilon_1 = \frac{1}{f_{1m}\overline{f_{1m}}}, \qquad \varepsilon_2 = \frac{1}{f_{2m}\overline{f_{2m}}}, \qquad \gamma = \frac{f_{1m}\overline{f_{1m}}}{f_{2m}\overline{f_{2m}}} = \frac{\varepsilon_2}{\varepsilon_1},$$

leads to

$$r_{1,2} = \frac{q}{2}\left(a_{11} + a_{22} \pm (a_{11} - a_{22})\sqrt{1 + \varepsilon\frac{4\gamma}{(1 + \varepsilon_2)^2\left(1 - \varepsilon\gamma\dfrac{1 + \varepsilon_1}{1 + \varepsilon_2}\right)^2}}\right).$$

If $|a_1| \gg |a_2|$, i.e., the modulus of the ε is small and the values $\varepsilon_1, \varepsilon_2,$ γ are finite numbers, the following estimations

$$r_1 \approx qa_1(1 + f_{2m}\overline{f_{2m}}) \left(1 + \frac{\varepsilon\gamma}{(1+\varepsilon_2)^2}\right),$$

$$r_2 \approx qa_2(1 + f_{1m}\overline{f_{1m}}) \left(1 - \frac{1}{(1+\varepsilon_1)(1+\varepsilon_2)}\right).$$

are valid.

We will estimate now the influence of the λ - transformation on the spectral conditionality number $K = \frac{|r_{max}|}{|r_{min}|}$.

For the nontransformed system (2.19) it is equal to $K_t = \frac{|a_1|}{|a_2|}$. For the transformed system (2.22) we have

$$K_\lambda = \frac{|r_1|}{|r_2|} \approx K_t \frac{1 + f_{2m}\overline{f_{2m}}}{1 + f_{1m}\overline{f_{1m}}} \left(\frac{1}{\left|1 - \frac{1}{(1+\varepsilon_1)(1+\varepsilon_2)}\right|}\right).$$

Another important spectral characteristic is the spectral scatter $S = \max_{i,j} |r_i - r_j|$. For the system (2.19) it takes the value $S_t = |a_1 - a_2|$ while the for system (2.22) it is

$$S_\lambda = |r_1 - r_2| \approx qS_t |\frac{1}{1-\varepsilon}(1 + f_{2m}\overline{f_{2m}}) - \frac{\varepsilon}{1-\varepsilon}(1 + f_{1m}\overline{f_{1m}})| \approx$$

$$\approx qS_t(1 + f_{2m}\overline{f_{2m}}).$$

Therefore, the spectral conditionality number can decrease or increase, but the spectral scatter is always decreased by λ - transformation.

We use now the results of Brauer [12] and Gershgorin [45] for estimation of eigenvalues of the matrix of the system (2.22) when $|a_1| > |a_2|$. We do not require that $|a_1| \gg |a_2|$. Brauer demonstrated that maximal eigenvalue r_{max} must satisfy the condition

$$|r_{max}| \leq \min(R, T),$$

where $R = \max_i R_i$, $T = \max_j T_j$; R_i, T_j are the sums of absolute values of elements of i - th row and j - th column of the matrix. In our case

$$R = q \max(|a_{11}| + |a_{12}|, |a_{21}| + |a_{22}|),$$

$$T = q \max(|a_{11}| + |a_{21}|, |a_{12}| + |a_{22}|)$$

and this inequality takes the form

$$|r_{max}| \leq q(|a_{11}| + |a_{12}|) = q|a_1|(1 + f_{2m}\overline{f_{2m}} + |\varepsilon f_{1m}\overline{f_{2m}}|),$$

or

$$|r_{max}| \leq q(|a_{21}| + |a_{22}|) = q|a_1|(|f_{2m}\overline{f_{1m}}| + |\varepsilon|(1 + f_{1m}\overline{f_{1m}})).$$

It follows from Gershgorin theorem that for the smallest eigenvalue the estimation is true

$$|r_{min}| \geq \min_i(|a_{ii}| - P_i),$$

where $P_i = R_i - |a_{ii}|$, a_{ii} is the diagonal element of the matrix. For the matrix of system (2.22) this estimation is of the form

$$|r_{min}| \geq q \min(|a_{11}| - |a_{12}|, \ |a_{22}| - |a_{21}|)$$

and we arrive to the inequality

$$|r_{min}| \geq q|a_2|(1 + f_{1m}\overline{f_{1m}} - |f_{2m}\overline{f_{1m}}|/|\varepsilon|), \qquad |a_{22}| - |a_{21}| > 0,$$

$$|r_{min}| \geq 0, \qquad |a_{22}| - |a_{21}| \leq 0.$$

Thus, the following estimations for the spectral conditionality number and spectral scatter take place

$$K_\lambda = \frac{|r_{max}|}{|r_{min}|} \approx K_t \frac{1 + f_{2m}\overline{f_{2m}}}{1 + f_{1m}\overline{f_{1m}} - |f_{2m}\overline{f_{1m}}|/|\varepsilon|}, \qquad |a_{22}| - |a_{21}| > 0,$$

$$S_\lambda = |r_{max} - r_{min}| \approx S_t q(1 + f_{2m}\overline{f_{2m}} + |\varepsilon f_{2m}\overline{f_{1m}}|).$$

Therefore, the above conclusions are true.

An inequality for the scatter S of the matrix A of order n is given in review of Marcus and Minc [80]

$$S \leq \left[2\left(1 - \frac{1}{n}\right)(tr A)^2 - 4E_2(A)\right]^{\frac{1}{2}}.$$

Here $tr(A)$ is the trace of the matrix A and $E_2(A)$ is sum of all principal second - order minors. For the matrix of system (2.22) for small ε this estimation leads to the inequality

$$S_\lambda \leq q S_t(1 + f_{2m}\overline{f_{2m}}),$$

for which the expression for the spectral scatter, obtained above, is a limiting case.

It follows from analysis of the obtained relation that λ - transformation improves the spectral characteristics of the system of equations, since it reduces the eigenvalues, the spectral conditionality number and, in general, the spectral scatter of the matrix.

This result can be illustrated better by means of the Girshgorin's theorem [45]. According to this theorem every eigenvalue r of the matrix $A = \|a_{ij}\|$ of order n is always located in one of following circles

$$|a_{ii} - r| \leq \sum_{j=1, j\neq i}^{n} |a_{ij}|, \qquad i = \overline{1, n}.$$

In our case this means that characteristic number r_1, r_2 of the matrix of the system (2.22) will lie within the circles

$$\left| q a_1 (1 + f_{2m}\overline{f_{2m}}) - r_1 \right| \leq q |a_2 f_{1m}\overline{f_{2m}}|,$$

$$\left| q a_2 (1 + f_{1m}\overline{f_{1m}}) - r_2 \right| \leq q |a_1 f_{2m}\overline{f_{1m}}|.$$

Fig. 2.5 shows the situation for the case of real characteristic roots. The intervals of localization of r_1 and r_2 are indicated by the brakes.

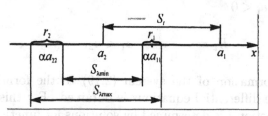

Figure 2.5.

We will demonstrate now that λ - transformation is useful for the system of equations (2.19), because the use of λ-parameter reduces the number of steps required by numerical solution starting from initial point $A_0(y_{10}, y_{20}, t_0)$ and ending at final point B (Fig. 2.6) comparably to that for the original argument t. In other words, we will prove that at any point of the integral curve the following inequality is true

$$H_\lambda \cos\theta \geq H_t. \tag{2.23}$$

Here θ is the angle between the tangent to the integral curve and t - axis, H_t and H_λ are minimal steps of integration with respect to the variables t and λ, correspondingly, for which the iterative process of Euler formula ceases to converge.

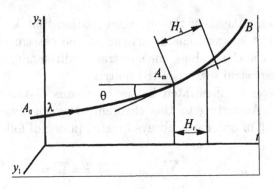

Figure 2.6.

The explicit Euler scheme for the system of equation (2.19) is of the form

$$y_{im+1} = y_{im} + h_t a_i y_{im} = (1 + h_t a_i) y_{im},$$

$$m = 1, 2, \ldots , \qquad i = 1, 2,$$

where h_t is the integration step with respect to t.

This scheme is absolutely stable if the condition $|1 + h_t a_i| < 1$ is fulfilled, i.e., if $a_i < 0$

$$H_t = -\frac{2}{a_1}, \qquad a_1 < a_2. \qquad (2.24)$$

After transformation of the system (2.19) to the form of (2.20) a system of three differential equations is obtained. For this system the solution for function t is determined by solutions for functions y_i. If the equations are linearized in the vicinity of a certain value $y_i = y_{im}$ the condition of stability of the transformed problem (2.22) is of the form (for the case $a_1 \ll a_2$)

$$|1 + h_\lambda r_{max}| < 1,$$

where h_λ is the step of integration with respect to λ. This inequality is satisfied if $a_i < 0$ and

$$H_\lambda = -\frac{2}{q a_1 (1 + f_{2m}^2)}. \qquad (2.25)$$

Using $\cos \theta = (1 + f_{1m}^2 + f_{2m}^2)^{-1/2}$ and taking into account (2.24) and (2.25) we obtain the inequality

$$\frac{H_\lambda \cos \theta}{H_t} = \frac{1 + f_{1m}^2 + f_{2m}^2}{1 + f_{2m}^2} \geq 1.$$

This relation proves the inequality (2.23) and consequently the effectiveness of the λ - transformation for the system ODE (2.19), i.e., for two - dimensional case.

We compare now the errors of the explicit Euler method before and after λ - transformation. We consider the solution of the Cauchy problem for one differential equation

$$\frac{dy}{dt} = f(y, t), \qquad y(t_0) = y_0, \qquad t \in [t_0, T]. \tag{2.26}$$

We assume that the solution of the problem is sufficiently smooth.

According to the Euler method the value of function y for the $(m+1)$ - th step is determined by the formula

$$y_{m+1} = y_m + h_t f(y_m, t_m), \qquad m = 0, 1, 2, \ldots ,$$

where h_t is the step of integration with respect to variable t, y_m is numerical solution for the m - th step.

The approximation error [81] of this formula at point $t = t_{m+1}$ can be represented as

$$\Delta_t = y(t_{m+1}) - y(t_m) - h_t f(y(t_m), t_m) =$$
$$= y(t_{m+1}) - y(t_m) - h_t \frac{dy(t_m)}{dt}.$$

Here $y(t_{m+1})$ is the exact solution at the point $t = t_{m+1}$.

Taking into account the representation of the function $y(t)$ by the Tailor formula in the vicinity of the point $t = t_m$ we obtain

$$\Delta_t = \frac{h_t^2}{2} \frac{d^2 y(\tau)}{dt^2}, \qquad t_m < \tau < t_{m+1},$$

where the right - hand side is Lagrange's remainder term of the Tailor formula.

After using of the λ - transformation the problem (2.26) becomes

$$\frac{dy}{d\lambda} = \frac{f(y, t)}{\sqrt{1 + f^2}}, \qquad \frac{dt}{d\lambda} = \frac{1}{\sqrt{1 + f^2}}, \tag{2.27}$$
$$y(0) = y_0, \quad t(0) = t_0, \quad \lambda \in [0, \Lambda].$$

The approximation error of the Euler formula for this problem at the point $\lambda = \lambda_{m+1}$ is equal to

$$\Delta_\lambda = y(\lambda_{m+1}) - y(\lambda_m) - h_\lambda F[y(\lambda_m), t(\lambda_m)] =$$

$$= y(\lambda_{m+1}) - y(\lambda_m) - h_\lambda \frac{dy(\lambda_m)}{d\lambda},$$

where h_λ is the step of integration with respect to λ, $F(y,\ t) = \dfrac{f(y,t)}{\sqrt{1+f^2}}$.

If the error of the right - hand side of this formula is estimated by the Lagrange's remainder then we obtain the equality

$$\Delta_\lambda = \frac{h_\lambda^2}{2} \frac{d^2 y(\nu)}{d\lambda^2}, \qquad \lambda_m < \nu < \lambda_{m+1}.$$

Taking into account the system of equations (2.26), (2.27) the second derivative of the formula can be written in the form

$$\frac{d^2 y}{d\lambda^2} = F_y \frac{dy}{d\lambda} + F_t \frac{dt}{d\lambda} = \frac{1}{\sqrt{1+f^2}}(F_y f + F_t) =$$

$$= \frac{f_y f + f_t}{(1+f^2)^2} = \frac{1}{(1+f^2)^2} \frac{d^2 y}{dt^2}.$$

Here a subscript indicates the variable with respect to which the differentiation is performed.

Let $t_m = t(\lambda_m)$. Then, if steps of integration are sufficiently small, it can be assumed that $\tau \approx t(\nu)$. Therefore,

$$\Delta_\lambda \approx \frac{h_\lambda^2}{h_t^2} \frac{\Delta_t}{(1+f^2)^2}.$$

It follows from the stability condition of the explicit Euler scheme that the steps of integration must satisfy the inequality (2.17), i.e.,

$$h_\lambda \cos\theta \geq h_t, \quad h_\lambda \geq h_t \sqrt{1+f^2}.$$

Since $\cos\theta = \dfrac{1}{\sqrt{1+f^2}}$ we obtain

$$\Delta_\lambda \leq \frac{\Delta_t}{1+f^2}. \tag{2.28}$$

It is shown in [81] that for explicit multistep linear schemes of integration of the problem (2.26) the total error of the method is equal to the approximation error up to a constant (equal to one for Euler method).

Therefore the inequality (2.28) shows that the total error of explicit Euler method becomes smaller than the error of traditional approach when λ - transformation is used. And the larger is the right - hand side of the equation (2.26) the bigger is the gain in the accuracy.

3. Algorithms, softwares, examples

Various approaches to the solution of the systems nonlinear algebraic or transcendental equations with a parameter were discussed in monograph [48]. It was pointed out that all solution continuation methods can be divided into two groups. The first group includes methods of discrete continuation based on iterative processes described usually by Newton - Raphson formula . The continuous methods form the second group. As a rule those are based on the Euler method and its modifications or Runge - Kutta method of integration of the Cauchy problem for ODE. In [48] the advantages and limitations of these two approaches were discussed. It was suggested there that apparently the combination of these two approaches is most effective. It was shown in the section 1.1 that both approaches are reduced to the integration of Cauchy problem with respect to a parameter by means of explicit and implicit schemes.

Thus, the methods of choice in developing the softwares for numerical integration of the Cauchy problem for ODE will be prediction and correction methods. These methods combine the advantages of explicit and implicit schemes. In addition the predictor-corrector methods are most appropriate for the integration of the stiff systems considered later.

The prediction will be implemented by explicit formulas and implicit formulas will be used for correction. That gives us an opportunity to design the iterative process based on discrete continuation.

The Euler method can be used as a starting point for the simplest algorithm. Then solving the problem (2.1) the first prediction is obtained according to the formula

$$y^p_{i\ m+1} = y_{i\ m} + h f_i(y_m, t_m),$$

$$i = \overline{1, n}; \qquad m = 1, 2, \ldots; \qquad y_m \in {}^n$$

$$y_m = (y_{1m}, \ldots, y_{nm})^T.$$

The first correction is calculated by the formula

$$y^c_{i\ m+1} = y_{i\ m} + h f_i(y^p_{m+1}, t_{m+1}), \tag{2.29}$$

where $y^p_{m+1} = (y^p_{1\ m+1}, \ldots, y^p_{n\ m+1})^T$.

The subsequent calculations using the formula (2.29). They are performed as a simple iteration process in which the corrected value obtained by the k - th step of iteration is used as a prediction for the following $(k + 1)$ - th step

$$y^{(k+1)}_{i\ m+1} = y_{i\ m} + h f_i(y^{(k)}_{m+1}, t_{m+1}), \qquad k = 1, 2, \ldots \tag{2.30}$$

The value δ

$$\delta = \sqrt{\sum_{i=1}^{n}(y_i^{(k+1)}{}_{m+1} - y_i^{(k)}{}_{m+1})^2}.$$

is used as a error on the $(m+1)$ - th step of integration of the problem (2.1).

If after 10 iterations the value of the error δ was still bigger than the preassigned accuracy ε the step of integration h was bisected and the computation process was repeated starting from the point (y_m, t_m) with the new value of step size. If the given accuracy ε was more then 100 times the error δ and iteration process defined by formula (2.30) was converged after the first iteration the step size was doubled. This approach is mentioned in [42] as advantageous for stiff systems. We developed the FORTRAN subroutine PC1 based on the algorithm described above.

Note that the problems (2.1) and (2.7) can be solved not only by means of the developed software PC1 but also with the help of any numerical method of integration of the Cauchy problem for ODE.

Now let us investigate the efficiency of λ - transformation applied to differential equations the right - hand sides of which can become infinite. It is obvious that the insurmountable difficulties can arise if such equations are solved by ordinary approaches.

Example 2.1. The cubic parabola.

Let us investigate the solution of the Cauchy problem

$$\frac{dv}{du} = \frac{1}{3v^2}, \qquad v(-1) = -1. \tag{2.31}$$

which has an exact solution $v = \sqrt[3]{u}$.

At the point $(0,0)$ on the integral curve of this problem the right - hand side of the equation becomes infinite. This point lies inside the interval $[-1,1]$ where the solution of the problem is sought. For parametric analysis of the solution we change to the new coordinate axes (x,y) rotated through an angle α with respect to the axes (u,v). The solid curve at Fig. 2.7 shows the solution of the problem (2.31) $v = \sqrt[3]{u}$. According to the relationship between new and old variables

$$y = v\cos\alpha - u\sin\alpha, \qquad x = v\sin\alpha + u\cos\alpha,$$
$$v = y\cos\alpha + x\sin\alpha, \qquad u = -y\sin\alpha + x\cos\alpha,$$

we rewrite the problem (2.31) in variables x, y

$$\frac{dy}{dx} = \frac{P(x,y)}{Q(x,y)}, \qquad y(-\sin\alpha - \cos\alpha) = -\cos\alpha + \sin\alpha, \tag{2.32}$$

$$P(x,y) = \cos \alpha - R(x,y) \sin \alpha,$$

$$Q(x,y) = R(x,y) \cos \alpha + \sin \alpha,$$

$$R(x,y) = 3(y \cos \alpha + x \sin \alpha)^2$$

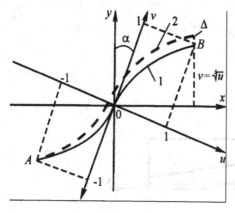

Figure 2.7.

The exact solution of this problem is given by the expression

$$y \cos \alpha + x \sin \alpha =$$
$$= \sqrt[3]{x \cos \alpha - y \sin \alpha}$$
(2.33)

As it is easily seen, the problem (2.32) converges to the problem (2.31) and the solution (2.33) converges to the solution $y = \sqrt[3]{x}$ when $\alpha \longrightarrow 0$.

After applying the λ - transformation the problem (2.32) takes the form

$$\frac{dy}{d\lambda} = \frac{P(x,y)}{Z(x,y)}; \qquad \frac{dx}{d\lambda} = \frac{Q(x,y)}{Z(x,y)}$$
$$y(0) = -\cos \alpha + \sin \alpha, \qquad x(0) = -\cos \alpha - \sin \alpha, \qquad (2.34)$$
$$Z(x,y) = \sqrt{P^2(x,y) + Q^2(x,y)}.$$

Note as Δ the following value

$$\Delta = \left| y^* \cos \alpha + x^* \sin \alpha - \sqrt[3]{x^* \cos \alpha - y^* \sin \alpha} \right|,$$

where x^*, y^* are values of x, y obtained as a result of numerical solution of problem (2.32) and (2.34), i.e., Δ is the modulus of the residual of equation (2.33).

Problems (2.32) and (2.34) were integrated by software PC1 and by Runge - Kutta method of 4 - th order. The results obtained by Runge - Kutta method are demonstrated below. At Fig. 2.8 the value of Δ is shown as a function of α for equal integration time of the problems in interval $x \in [-\sin \alpha - \cos \alpha, \sin \alpha + \cos \alpha]$. This time corresponds to the integration step $h_x = 0.25/40$ for problem (2.32) and step $h_\lambda = 0.25/20$ for problem (2.34). Curves 1 and 2 were obtained as solutions of problems (2.32) and (2.34) respectively.

At Fig. 2.9 the value Δ is given as a function of dimensionless computing time t/t_0 for $\alpha = 1/1024$. Curves 1 and 2 also correspond to problems (2.32) and (2.34).

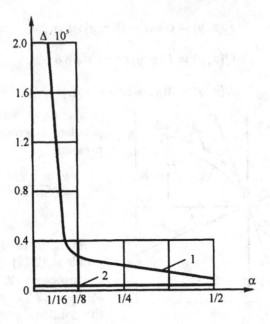

Figure 2.8.

From analysis of these results it follows that:

— if the λ - transformation is used the error Δ does not depend on the angle α (curve 2, at Fig. 2.8) whereas the accuracy substantially depends on α (curve 1) for initial equation;

— the error of integration of the problem (2.32) is a nonmonotonic function of the integration step and/or calculation time t in turn (curve 1, at Fig. 2.9). The local maximum at this curve corresponds to the fact that for some integration step some of the solution points are close to the point $(0, 0)$ where the right - hand side of the equation is large. The curve 2 shows monotonic increase of the error of the λ−transformed problem. This apparently results from the increase of rounding error due to a reduction in the integration step;

— the full error Δ of the initial problem (2.32) can not be lesser than 10^{-4} by any integration step h_x whereas for the transformed problem (2.34) the error reaches the minimal value $0.18 \cdot 10^{-6}$.

We also used the software DLSODE from solver ODEPACK [57] for the solution of this problem. The solution of (2.32) was obtained for low accuracy ($\varepsilon_{abs} = 10^{-4}$). For large step corresponding to that low accuracy and the the singular point is missed by the program. This is no longer the case if accuracy is increased. High accuracy $\varepsilon_{abs} = 10^{-8}$ leads to the increase of error and overflow. The solution of λ-transformed problem (2.34) was successful in every case.

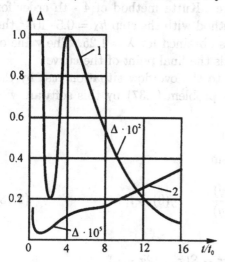

Figure 2.9.

Example 2.2. The Brachistochrone.

The problem of the brachistochrone (the steepest descent curve) was considered by J. Bernoully in 1696. The problem is to integrate the following differential equation

$$\frac{dy}{dx} = \sqrt{\frac{L}{y} - 1}, \quad (2.35)$$

where L is some positive constant.

If the solution of the equation with initial condition

$$y(0) = 0, \quad (2.36)$$

is obtained the right - hand side becomes infinitely large by $x = 0$ and majority of numerical methods of ODE integration fails. We will obtain the solution of the problem (2.35), (2.36) for $y \in [0, L]$. The λ - transformation reduces the problem to the following

$$\frac{dy}{d\lambda} = \sqrt{\frac{L-y}{L}}, \quad \frac{dx}{d\lambda} = \sqrt{\frac{y}{L}}, \quad y(0) = 0, \quad x(0) = 0 \quad (2.37)$$

To estimate the accuracy of the numerical solution we find an analytical solution that can be represented as

$$x = R(t - \sin t), \qquad y = R(1 - \cos t), \qquad R = \frac{L}{2}.$$

If the parameter t is eliminated we obtain

$$\cos t = 1 - \frac{y}{R}, \qquad t = \frac{x}{R} + \sqrt{1 - \left(1 - \frac{y}{R}\right)^2}.$$

Therefore we use the following expression for estimating the accuracy

$$\Delta = \left| \frac{y^*}{R} - 1 + \cos \left(\frac{x^*}{R} + \sqrt{1 - \left(1 - \frac{y^*}{R}\right)^2} \right) \right|,$$

where x^* and y^* are numerical values of x and y.

The problem (2.37) was integrated equally successful both by the software PC1 and by using Runge - Kutta method of 4 - th order for $L = 2$. For the Runge - Kutta method with the step $h_\lambda = 0.5 \cdot 10^{-2}$ the maximal value $\Delta = 0.8 \cdot 10^{-4}$ was obtained for $\lambda = 1.25$. The value of the argument $\lambda = 3.75$ corresponds the final point of the curve.

The software DLSODE get in to the overflow situation just by the first step. But integration of the problem (2.37) by this software was successful.

Example 2.3. The Pascal spiral.

We consider the Cauchy problem

$$\frac{dy}{dx} = \frac{P(x,y)}{Q(x,y)}, \qquad y(0) = l, \qquad (2.38)$$

where

$$P(x,y) = lx^2 - S(x,y)(2x - a),$$

$$Q(x,y) = y(2S(x,y) - l^2),$$

$$S(x,y) = x^2 + y^2 - ax,$$

a and l are given numbers. The solution of the problem is a closed curve named as Pascal spiral. The form of the curve is shown in Fig. 2.10 for the case when

$$a < l < 2a.$$

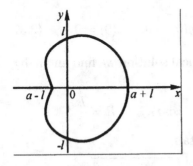

Figure 2.10.

The Pascal spiral is described by the equation

$$S^2(x,y) - l^2(x^2 + y^2) = 0.$$

Hence, the accuracy of computations can be estimated by the expression

$$\Delta = |S^2(x^*,y^*) - l^2(x^{*2} + y^{*2})|,$$

where x^*, y^* are numerical values of x and y.

Ordinary numerical methods and the software DLSODE among them are unable to construct the whole integration curve of the problem (2.38).

However, this problem can be solved by software DLSODE, PC1, Runge - Kutta method and other numerical method if the λ - transformation is used. Then the problem is reduced to the form

$$\frac{dy}{d\lambda} = \frac{P(x,y)}{Z(x,y)}; \quad \frac{dx}{d\lambda} = \frac{Q(x,y)}{Z(x,y)}, \qquad y(0) = l, \quad x(0) = 0, \qquad (2.39)$$

where $Z(x,y) = \sqrt{P^2(x,y) + Q^2(x,y)}$.

This problem was solved for $a = 1$, $l = 1.5$ by software DLSODE, PC1, and Runge - Kutta method of 4 - th order. The parameter λ was changed from 0 to 10.5. The value of Δ was not bigger than $0.2 \cdot 10^{-4}$.

The illustrative examples considered above are artificial. But the same problems arise in the solution of real problems. Consider for example the Euler's kinematic equations [76]

$$\begin{pmatrix} \omega_1 \\ \omega_2 \\ \omega_3 \end{pmatrix} = \begin{pmatrix} \sin\theta\sin\varphi & 0 & \cos\varphi \\ \sin\theta\cos\varphi & 0 & -\sin\varphi \\ \cos\theta & 1 & 0 \end{pmatrix} \begin{pmatrix} \psi_{,t} \\ \varphi_{,t} \\ \theta_{,t} \end{pmatrix}, \qquad (2.40)$$

that determine relation between vector of angular velocity $\omega(\omega_1(t),\ \omega_2(t),\ \omega_3(t))$ in the body fixed coordinate system and time derivatives of Euler's angles ψ, φ, θ (ψ is the angle of precession, φ is the angle of eigenrotation and θ is the angle of nutation).

To determine Euler's angles as a functions of time we resolve the system with respect to derivatives and obtain [76]

$$\frac{d\psi}{dt} = \frac{S(\varphi)}{\sin\theta},$$

$$\frac{d\varphi}{dt} = \omega_3 - S(\varphi)\operatorname{ctg}\theta, \qquad (2.41)$$

$$\frac{d\theta}{dt} = \omega_1\cos\varphi - \omega_2\sin\varphi,$$

$$S(\varphi) = \omega_1\sin\varphi + \omega_2\cos\varphi$$

It is obvious that numerical integration for angle of nutation $\theta = 0$ is difficult, because in this case the right - hand sides of the first two equations become infinite. But this problem vanishes if the λ - transformation is applied to the system (2.41). Then the system can be represented in the form

$$\frac{d\psi}{d\lambda} = \frac{S(\varphi)}{Q} \operatorname{sign}(\sin\theta)$$

$$\frac{d\varphi}{d\lambda} = \frac{\omega_3 \sin\theta - S(\varphi)\cos\theta}{Q} \operatorname{sign}(\sin\theta)$$

$$\frac{d\theta}{d\lambda} = \frac{\omega_1 \cos\varphi - \omega_2 \sin\varphi}{Q} |\sin\theta| \tag{2.42}$$

$$\frac{dt}{d\lambda} = \frac{|\sin\theta|}{Q}$$

$$Q = \Big[\sin^2\theta + S^2(\varphi) + \sin^2\theta(\omega_1 \cos\varphi - \omega_2 \sin\varphi)^2 +$$

$$+ (\omega_3 \sin\theta - S(\varphi)\cos\theta)^2\Big]^{\frac{1}{2}},$$

which can be integrated for the $\theta = 0$ too.

The system of equations (2.42) was solved numerically by PC1 software for $\omega_1 = -1$, $\omega_2 = 1$, $\omega_3 = 0$ and for the initial conditions $\lambda = 0$, $\psi = \varphi = \theta = t = 0$. The obtained solution determined the relation between the Euler angles of the form $\psi = -\varphi$.

Among many other practical cases, we only mention similar problems arising in the numerical integration of the kinematic equations for the airplane angles [73]. This equations have a singularity when the pitch angle is equal to $\pi/2$ and can be solved efficiently with the help of λ-transformation.

Chapter 3

STIFF SYSTEMS OF ORDINARY DIFFERENTIAL EQUATIONS

The experience of numerical integration of Cauchy problems for ordinary differential equations shows that it is especially the stiff systems that require a special numerical methods. In this chapter the effect of λ-transformations on such systems of ordinary differential equations and on methods of their integration will be considered.

1. Characteristic features of numerical integration of stiff system of ordinary differential equations

Many problems in aerodynamics, ballistics, dynamics and control of aircrafts and rockets, chemical kinetics, kinetic of elementary processes of atomic, molecular and nuclear physics ets. are reduced to numerical integration of Cauchy problem for system of ordinary differential equations. For simulating these processes, it is necessary to take into account a large number of parameters. For perfect description of processes in any interval of observation this often requires using two types of decreasing functions: fast and slow. Since first of them decrease fast, hence only slow decreasing functions can be observed on the time scale of the whole processes. Nevertheless, in any moment of the observation a possibility for appearing of the fast decreasing process remains.

Such phenomenon is named as stiffness and the systems of ordinary differential equations implemented such processes are named as stiff systems.

Note, that stiffness of problem is a property of the mathematical model rather than a feature of numerical method applied. Mathematical

67

stiffness of problem reflects the fact that in considered physical model various processes have different rates.

A characteristic feature of all stiff system is that the solution of their Cauchy problem either have fast initial variations or large variations in certain interval of observation (in a boundary layer).

Such a type of equations is considered separately due to the difficulties of their numerical integration by classic methods, for example, by explicit onestep and multistep methods. It became clear that if small integration step is used for implementation of fast processes in boundary layer, it can not be increased outside of the boundary layer, although derivatives become essentially smaller. Even a very small increase of a certain value of the step, determined by the method used and by equation solved, leads to a sharp increasing of error. Indeed, as has been shown in [7, 95, 81, 102], for example, in order to provide the absolute stability of the numerical solution of the Cauchy problem

$$\frac{dy}{dt} = f(y, t), \qquad y(t_0) = y_0, \qquad t \in [t_0, \, T] \tag{3.1}$$

$$y = (y_1, \, \cdots, \, y_n)^T, \qquad f = (f_1, \, \cdots, \, f_n)^T,$$
$$y_0 = (y_{10}, \, \cdots, \, y_{n0})^T,$$

it is necessary to use such step of integration h, for which each of the values $H_i = h\lambda_i$ ($i = \overline{1, \, n}$) would remain within the domain of absolute stability. Here λ_i is an eigenvalue of Jacoby matrix $\partial f / \partial y$ which can in general be complex. Thus, for methods with the bounded stability domain the step length is limited by order of magnitude of the minimal time (or length) constant of the system. Since the integration interval can be much more than that constant, the required number of integration steps can be extreme large.

In order to eliminate these limitations, the numerical methods were proposed (see, for example, [42, 16, 109]) that allow substantial increase of the integration step outside of the boundary layer. However, the problem of numerical solution of stiff systems remains difficult [4, 53, 74, 34, 75, 35].

Initially stiff systems were viewed as a special and exotic case, which simply required some extra computational power. However, when a number of problems was solved with the help of various numerical methods using more and more powerful computers it became clear that stiffness is rather a rule than an exception.

It is known that careless neglecting of "small values" in mathematical simulation of real processes can substantially distort the situation. Therefore, in systems of equations modeling complicated processes it is often necessary to take into account a large number of factors which seem unimportant at the first sight. This often results in a large order of the system of equations and its stiffness.

It should be also noted that hidden forms of stiffness exist too. For example, a large class of smooth finite - dimensional optimization problems [95, 124] is difficult to solve by traditional methods because the level surfaces have a shape of a valley. Investigation of this phenomena [95] shows that these difficulties are connected with stiffness of the system of differential equations, describing the trajectory of steepest descent. This fact has to be taken into account in construction of optimization algorithms in these cases.

Note, that problem described by systems with partial derivatives can be stiff if their solution is reduced to systems of ordinary differential equations. It is shown in [16] that the problem of heat conductivity for a rod is a stiff one. The problem of an elastic beam movement was pointed out to be stiff in [53].

We note the following characteristics of stiff linear systems [95]:

1. Almost always there are two regions with different behavior of the solution. We say "almost always" because the initial conditions can be chosen such that the boundary layer is eliminated. However, even in this case the stiffness of the equations is not eliminated.

2. The eigenvalues λ_i of the Jacoby matrix $J = \partial f / \partial y$ completely determine the solution.

3. The stiffness of homogeneous system follows from stiffness of non-homogeneous one

$$\frac{dy}{dt} = Jy + g(t), \qquad y \in {}^n, \qquad t \in [t_0, \, T] \qquad (3.2)$$

This confirms the fact that the stiffness is an interior property of the system and it can not appear only because of variations of the function $g(t)$.

4. Outside of the boundary layer linear relations between components of vector function $y(t)$ can be determined. The number of these relations is equal to the number of fast oscillating particular solutions of the system (3.2). I.e., outside of the boundary layer the solution of a stiff system can be described by solution of a smaller order system, which is not stiff (see, for example, [42]).

Despite a large number of publications related to this problem a general concept of stiff system does not exist [7, 34, 35]. Furthermore, there is no generally accepted definition of stiffness.

In [109] the definition of stiffness is connected with the concept of stability. A system is considered to be stiff if it does not have unstable component of solution but has several very stable components. The Jacoby matrix of such system does not have eigenvalues with large positive real parts and has at least one eigenvalue with large negative real part.

The following definition is used in [72, 102].

The Cauchy problem (3.1) is called stiff on a certain interval $I \subset [t_0, T]$ if for $t \in I$

$$1) \ Re \ \lambda_i(t) < 0, \qquad i = \overline{1, n},$$

$$2) \ S(t) = \frac{\max\limits_{1 \leq i \leq n} |Re \ \lambda_i(t)|}{\min\limits_{1 \leq i \leq n} |Re \ \lambda_i(t)|} \gg 1, \qquad (3.3)$$

where λ_i are the eigenvalues of the Jacoby matrix $J = \partial f / \partial y$ in which $y = y(t)$ is a solution of (3.1).

In [95] a detailed critical discussion of most well - known definitions is given. The authors themselves recommend the following definition: the system of ordinary differential equations

$$\frac{dy}{dt} = f(y, t), \qquad f(y, \ t) \in C_{ty}^{(d,d)}(\Gamma)$$

$$\Gamma \subset I_t \times \frac{n}{y}, \qquad I_t = \{0 \leq t \leq \infty\}$$

$$(3.4)$$

is called stiff on the interval of variation of argument $[a, \ b]$ belonging to interval of solution existence if for any vector of initial values $(y_0, t_0) \in \Gamma$ and any interval $[t_0, \ t_0 + T] \subset [a, \ b]$ exist τ, L, N which satisfy the following inequalities

$$\tau \ll b - a \qquad (3.5)$$

and

$$0 < L \leq \rho \left(\frac{\partial f}{\partial y}\right) \leq \left\|\frac{\partial f}{\partial y}\right\|, \qquad (y, \ t) \in \Gamma,$$

and make the following inequality

$$\left|\frac{dy_k}{dt}\right| \leq \frac{L}{N} \max_{t \in [t_0, \ t_0 + T]} |y_k(t)|, \qquad k = \overline{1, n}$$

$$t_0 + \tau \leq t \leq t_0 + T, \qquad N \gg 1.$$

valid. Here $\rho(\partial f / \partial y)$ is a maximal modulus of eigenvalues of Jacoby matrix (the spectral radius); $\| \cdot \|$ is a norm of the matrix.

If initial conditions are such that the boundary layer exists explicitly the value of N gives an order of magnitude of the ratio of derivatives within and outside the boundary layer, thus describing the decrease of derivatives after passing it.

The authors believe the nonseparable connection of system (3.4) stiffness with the interval of solution observation $[a, b]$, included in inequality (3.5), to be an advantage of their definition. If the system is considered only on the interval $[a, c] \subset [a, b]$, which includes only the boundary layer $\tau = c-a$, it can not be viewed as stiff on $[a, c]$ because no variation of the solution behavior is observed .

The definition of stiff system of equations introduced in [74, 75] is as follows.

Let $J(t) = J(y, t) = \|\partial f_i / \partial y_i\|$, $(i, j = \overline{1, n})$ to be the Jacoby matrix of function $f(y, t)$ where $y(t)$ is a solution of the problem (3.1).

In the neighborhood of a point (y_0, t_0) the following system of linear differential equations

$$\frac{dy}{dt} = J(t_0)(y - y_0) + f(y_0, t), \qquad (3.6)$$

corresponds to the system (3.1). It is a linearization with respect to y of the original system (3.1) in the neighborhood of (y_0, t_0). It is assumed, that the solution of the system of equation (3.6) for initial conditions $y(t_0) = y_0$ approximates the solution of Cauchy problem (3.1) for $t_0 \leq t \leq \tau$ where $0 \leq \tau \leq T$ and τ is a value, depending on t_0. Let $\lambda_i(t)$ $(i = \overline{1, n})$ to be the eigenvalues of the matrix $J(t)$ and all of them are real. Let eigenvectors $\xi_i(t)$ $(i = \overline{1, n})$ correspond to $\lambda_i(t)$. The spectrum of matrix J gives a correct information about qualitative behavior of system (3.1).

The Cauchy problem (3.1) is called stiff if an independent on t constant $C > 0$ exists such that

$$\lambda_i(t) \leq C \text{ for } t \in [t_0, T], \qquad i = \overline{1, n} \qquad (3.7)$$

and for these conditions the following inequalities

$$M = M(t) = \max(-\lambda_i(t)) > 0, \qquad S = M(T - t_0) \gg 1,$$
$$(3.8)$$
$$t \in [t_0, T].$$

take place.

In [34, 35] the Cauchy problem (3.1) is called stiff if the spectrum of the Jacoby matrix of system (3.1) can be divided on two essentially different parts:

1) the stiff spectrum where eigenvalues and eigenvectors are noted as $\Lambda_i(t)$ and $\Phi_i(t)(i = 1, \dots, I)$, correspondingly. The following conditions

are satisfied for this stiff spectrum

$$Re\,\Lambda_i(t) \leq -L < 0, \qquad |\,Im\,\Lambda_i(t)| < |\,Re\,\Lambda_i(t)|, \qquad i = \overline{1,\,I};$$

2) the soft spectrum where eigenvalues and eigenvectors are noted as $\lambda_j(t)$ and $\varphi_j(t)$ $(j = \overline{1,\,J};\ J = n - I)$. The following conditions are correct for this part of spectrum

$$|\lambda_j(t)| \leq l, \qquad j = \overline{1,\,J}$$

Also, the ratio L/l must be large.

Moreover, if the Jacoby matrix $\partial f / \partial y$ depends on t, it should be taken into account that, in general, $I = I(t)$ and $J = J(t)$.

Methods and algorithms used for the solution of stiff systems are reviewed in [42, 95, 72, 81, 53]. The computation programs for them are given in [16, 109, 53].

The term "stiff system" was apparently introduced in [22] where the problem

$$y' = -50(y - \cos t), \qquad y(0) = 0$$

was solved by methods Adams and Runge - Kutta. The efficiency of implicit methods for solving such equation was shown there.

Originally the phenomenon of instability with integration step limitations for hyperbolic system of partial differential equations was considered in the famous paper [19].

However, only in [23] the stiff system of ordinary differential equations were considered. Numerical instability was shown to be the reason of difficulties and general concepts and definitions were introduced.

To ensure the stability of numerical solution of the problem (3.1) it is necessary to choose such integration step h for which each of the (generally speaking complex) values $H_i = h\lambda_i$ $(i = \overline{1,\,n})$ would be inside of the absolute stability domain (λ_i is eigenvalue of the Jacoby matrix $J(t) = \dfrac{\partial f}{\partial y}$). Thus, for methods with bounded stability domain the length of the step is limited by the value of the minimal eigenvalue of the matrix J. The interval of integration can be mach more than $1/\min\limits_{1\leq i\leq m} |\lambda_i|$. As a result, the required number of integration steps can be comparable with the coefficient $s(t)$ given by the formula (3.3).

Indeed, for each integration step the system of equations (3.1) can be transformed into the linear system (3.2) with the Jacoby matrix. Assumed that all eigenvalues of this Jacoby matrix are different and making the corresponding change of variables, the matrix can be reduced to the diagonal form with eigenvalues $\lambda_i\ i = (\overline{1,\,n})$ on the main diagonal.

The system of equations (3.2) takes the form

$$\frac{dy_i}{dt} = \lambda_i y_i + e(t), \qquad \overline{1, n}$$

Because of that, the methods are usually tested on the equation

$$\frac{dy}{dt} = \lambda y, \qquad (3.9)$$

where $y(t)$ is a scalar function. The solution of this equation is stable asymptotically if $Re\,\lambda < 0$, it is unstable if $Re\,\lambda > 0$ and stable if $Re\,\lambda = 0$.

Numerical methods replace differential equation by a finite differences. This finite differences equations are stable asymptotically if modulus of all roots of its characteristic equation are smaller than unit. It is unstable if there is at least one root with its modulus of equals to unit. And it is stable if there are some roots with unit modulus and the for the rest the modulus is smaller than unit. Clearly, the methods should be used with such integration step (depending on λ) which ensures correspondence between the test differential equation and its finite differences equation for all types of stability.

The set of the values $h\lambda$ that satisfy the condition of asymptotic stability of the solution of the finite differences equation for test equation (3.9) by a numerical method is called to the stability (absolute stability) domain of the method in complex plane $h\lambda$.

It was shown in chapter 2 that stability domain of explicit Euler method is the interior of the unit circle with the center at the point $h\,Im\,\lambda = 0$, $h\,Re\,\lambda = -1$.

Analysis of stability domain of traditional algorithms of the Runge - Kutta and Adams shows that they are bad for solving stiff systems. A very small values of h required by these methods result in a huge number of computations. In connection with that, it was suggested in [23] that methods for which the stability domain of the test equation (3.9) is the whole left half - plane of the plane $h\lambda$ should be designed and applied for stiff systems. In other words for $Re\,\lambda < 0$ the solution of corresponding difference equation should be stable for any positive value of h. Such methods are called A - stable.

It was also proved in [23] that explicit linear multistep method can not be A - stable, and that the order of implicit multistep method can not be bigger than two. Thus, the majority of successful algorithms for solving are implicit algorithms. However, there are exceptions to this rule. It is show in [74, 75], how the stiff systems can be solved by the explicit Euler method. The proposed approach is based on the achievements of the theory of numerical stability for finite differences schemes and Chebyshev's iteration processes. On the base of this approach the program DUMKA is designed.

It was shown later that less limiting requirements are more useful for multistep methods. The concepts of $A(\alpha)$ - stability [123] and stiff stability [42] were introduced.

A numerical method is called to be $A(\alpha)$ - stable, $\alpha \in (0, \pi/2)$, if its stability domain includes the infinite wedge $|arg(-\lambda)| < \alpha$.

Previous definitions were talking only about stability. A more complex concept of stiff stability which includes both the stability and accuracy of approximation of the exponential fundamental solution of equation (3.9) is introduced in [42].

A method is called to be stiff stable if it is absolute stable in domain R_1 and precise in domain R_2 where $R_1 : \{Re(\lambda h) \leq D\}$, $R_2 : \{D < Re(\lambda h) \leq \alpha, |Im(\lambda h)| < \theta\}$, D, α, θ — are numbers.

The formal methodology of solution of stiff ordinary differential equations, using the back differentiation formulas, was described in [22]. First package of programs DIFSUB applying these formulas, which were not very popular before, was designed by Gear and described in [42]. This package allows to apply the modified Adams method to solution of unstiff systems of equations. Next package of programs was called STIFF (see [42]). We consider here the main ideas on which these programs are based.

The back differentiation formulas (BDF) of k - th order for on the $m + 1$ -th step of the solution of the equations system (3.1) have the form

$$y_{m+1} - hbf(y_{m+1}, t_{m+1}) - \sum_{i=0}^{k-1} a_i y_{m-i} = 0 \qquad (3.10)$$

For $k > 6$ the method described by these formulas are unstable. The Newton - Raphson method is used to solve the system of nonlinear equations (3.10)

$$y_{m+1}^{(j+1)} = y_{m+1}^{(j)} - \frac{F(y_{m+1}^{(j)}, t_{m+1})}{J_F(y_{m+1}^{(j)}, t_{m+1})}, \qquad (3.11)$$

where j is number of iteration

$$F(y_{m+1}^{(j)}, t_{m+1}) = y_{m+1}^{(j)} - hbf(y_{m+1}^{(j)}, t_{m+1}) - \sum_{i=0}^{k-1} a_i y_{m-i}^{(j)},$$

$$J_F(y_{m+1}^{(j)}, t_{m+1}) = \left. \frac{\partial F(y, t)}{\partial y} \right|_{y=y_{m+1}^{(j)}, t=t_{m+1}}.$$

Initial approximation is given by the polynomial approximation of unknown functions. It follows from formulas (3.10),(3.11) that the considered methods are methods of prediction - correction. To optimize application of the methods, an algorithm which controls the order of the method and integration step was developed. Subsequent improvement of these program were achieved by Hindmarsh and coworkers [110]. The package of programs GEAR developed in [110] is a modification of DIFSUB. It is a new method of solution accuracy evaluation (only local errors are checked), an improved method for calculation of Jacobian, and several modifications of the program.

The programpackege EPISODE [15] is intended to solve both stiff (by methods of back differentiation) and unstiff (Adams type methods) systems of ordinary differential equations. This package differs from the program package STIFF by methods of the integration step variation, the frequency of computation of the Jacobian, and in the local error estimation.

The most recent known to us version of Gear's programs is called LSODE [57, 56]. It applies the implicit formulas of Adams - Multon to solution of the unstiff differential equations. Stiff equations are solved by means of back differentiation formulas. In the iteration process (3.11) the modified Newton -Raphson method is used. Initial approximation is determined by explicit formulas that are analogous to (3.10). Jacobian is computed only if the convergence speed of the iteration process is small.

Another well - known package of programs for solution of stiff system is the package FACSIMILE [21]. It is based on back differentiation formulas and is intended to solve stiff problems of chemical kinetics.

The programs for solving stiff systems from program library of Computation Center of Moscow State University [120] and programs RKF4RW [96], RAI4 [14] should be mentioned as well. They allow to determine whether the problem is stiff or not. They are based on formulas Runge - Kutta. Moreover, the latter program uses an adaptive formulas. The program RADAU5 [53] also should be mentioned .

Let us consider peculiarities of applying the λ - transformation to stiff systems of equations (3.1) if linear multistep methods are used. The general form of representation of these methods is

$$\sum_{i=0}^{k} \alpha_i y_{m+i} = h \sum_{i=o}^{k} \beta_i f_{m+i}, \qquad m = 0,\ 1,\ 2,\ \ldots \qquad (3.12)$$

where α_i, β_i are constants, $\alpha_k \neq 0$, $|\alpha_0| + |\beta_0| \neq 0$.

Formulas (3.12) determine linear relations between y_{m+i} and f_{m+i}, $i = \overline{0,\ k}$. To compute the sequence of values y_m, it is necessary first to obtain somehow k initial values of y_0, y_1, $\ldots$, y_{k-1}. If $\beta_k \neq 0$,

the right - hand side of formula (3.12)contains $f_{m+k} = f(y_{m+k}, t_{m+k})$
and, in general case, to determine y_{m+k} it is necessary to solve nonlinear
equation. In this case method (3.12) is called implicit multistep one.
In comparison with the explicit methods [81], the implicit methods are
more accurate and stable because for perturbation of y_m the integration
step h can be larger for these methods.

It was shown previously that implicit methods are preferable for solution of stiff systems of equations. In this case, the problem is to find
a solution of the equation

$$y_{m+k} = \frac{h\beta_k}{\alpha_k} f(y_{m+k}, t_{m+k}) + g_m, \qquad \beta_k \neq 0, \qquad (3.13)$$

where function g_m includes the known values y_{m+j}, f_{m+j}, $j = \overline{0, k-1}$,
determined at previous steps of integration process for equations (3.1).
It is proved in [55] that if the integration step

$$h < \left| \frac{\alpha_k}{\beta_k} \right| \frac{1}{L}, \qquad (3.14)$$

where L is Lipschitz's constant ($\|f(y_1, t) - f(y_2, t)\| \leq L\|y_1 - y_2\|$ for
all $t \in [t_0, T]$) then the unique solution y_{m+k} of equation (3.13) exists.
It can be obtained by means of the iteration process

$$y_{m+k}^{(j+1)} = \frac{h\beta_k}{\alpha_k} f(y_{m+k}^{(j)}, t_{m+k}) + g_m, \qquad j = 0, 1, 2, \ldots \qquad (3.15)$$

It is important to note that the convergence speed depends on the
value of $h|\beta_k/\alpha_k|L$. The smaller is the value (comparably to unit) the
faster is the convergence. Such a requirement imposes strong restrictions
on the value of integration step h. That is why the Newton - Raphson
method (3.11) is used instead of simple iteration (3.15).

The situation improves if λ - transformation is applied. Let us consider
the case $n = 1$ for system (3.1) as an example. Let us suppose that
$L = |\frac{\partial f}{\partial y}| = |f_{,y}|$. After applying of λ - transformation to equation (3.1),
the first equation of the obtained system takes the form

$$\frac{dy}{d\lambda} = \frac{f(y, t)}{\sqrt{1 + f^2}}.$$

The Lipschitz's constant for the transformed equation is equal to

$$L_\lambda = \frac{|f_{,y}|}{|1 + f^2|^{\frac{3}{2}}} = \frac{L}{|1 + f^2|^{\frac{3}{2}}}.$$

Clearly that for large value of the right-hand side $f(y, t)$ the condition $L_\lambda \ll L$ is satisfied and the condition (3.14) is weaker. If we assume that $h_\lambda = h(1 + f^2)^{1/2}$, the inequality (3.14) takes the form

$$h < \left| \frac{\alpha_k}{\beta_k} \right| \frac{1 + f^2}{L}.$$

Examples given below confirm this. They show that in this case the modified method of simple iteration can be used.

The results of chapter 2 show that λ - transformation increases the stability domain of explicit Euler method and instability domain of implicit scheme. Also after applying of λ - transformation, the A - stability of the method remains. It was established, that this is very important for solution of stiff systems of equations.

Application of λ - transformation to the test system of equations (2.16) with eigenvalues of the system matrix satisfying the condition $|a_1| \gg |a_2|$ demonstrated that, if the orders of solutions y_1 and y_2 in formulas (2.20) are assumed to be equal, the spectral conditionality number of the transformed system is smaller than that of original system. As we saw above, this characteristic can be used as an indirect measure of system stiffness. The same is true with respect to the spectral scatter. As it is seen from (2.20), it decreases after applying of the λ - transformation. This fact shows that the eigenvalues of the transformed system get closer.

Thus, we come to the conclusion that λ - transformation smooths the stiffness of original system of equations and, therefore, increases the effectiveness of any numerical method of its solution.

2. Singular perturbed equations

Equations with a small parameter by highest derivative are called singular perturbed equations. They form a class of stiff systems convenient for studying the effectiveness of various numerical methods of integration of stiff systems. New advances in asymptotic theory [119], theory of difference schemes [28] and simplicity of qualitative behavior of solution make possible the detailed analyses of such a system.

We consider singular perturbed system of equations [35]

$$\dot{x} = f.$$

Let the system to be represented in the form

$$
\begin{aligned}
\dot{y} &= Y(y, z) \\
\dot{z} &= L\, Z(y, z) \qquad (\varepsilon \dot{z} = Z(y, z),\ \varepsilon = \frac{1}{L} \ll 1)
\end{aligned}
\tag{3.16}
$$

Here Y, Z are smooth vector functions of the order of $O(1)$, with the same order of their derivatives. Dimensions of vectors y, z are equal to k and l, respectively. Let us introduce the following notations for vectors: $x = (y, z)$, $f = (Y, Z)$.

The spectrum of the matrix $f_{,x}$ is determined by the equation

$$\begin{vmatrix} Y_{,y} - \lambda E_k & Y_{,z} \\ L Z_{,y} & L Z_{,z} - \lambda E_l \end{vmatrix} = 0,$$

where E_k and E_l are unit matrices of order k and l.

Clearly that stiff spectrum depends on the spectrum of matrix $L Z_{,z}$ and respective eigenvectors, having z as their determining components. Their components y are of the order $O(\frac{1}{L})$. The qualitative structure of the trajectory is well - known and is determined by the surface

$$\Gamma = \Big\{ x = \{y, z\} : \ Z(y, z) = 0 \Big\},$$

which divides the phase space into two parts: $Z(y, z) > 0$ and $Z(y, z) < 0$. System (3.16) is stiff only for those points $x(t)$ where spectrum of matrix $Z_{,z}$ is stable, i.e., where the real parts of eigenvalues are negative. A typical trajectory of the system (3.16) for scalar functions y and z is shown in Fig. 3.1. In general case, the picture is the same. Outside of a small neighborhood of the surface Γ the phase velocity has almost horizontal direction and very large absolute value ($\|\dot{x}\| = O(L)$, $\|\dot{y}\| = O(1)$, $\|\dot{z}\| = (L)$). If $Z(y, z) > 0$ the point $x(t)$ moves fast to the right and it moves to the left if $Z(y, z) < 0$. In short time, which order is $O(\frac{1}{L})$, the point $x(t)$ moves away from initial position x_0 in $O(\frac{1}{L})$ - neighborhood of surface Γ. Here $\dot{x} = O(1)$ and point $x(t)$ moves slow along the surface Γ. The surface is divided into two parts in accordance with the sign of $Z_{,z}$: a stable part of Γ ($Z_{,z} < 0$ in scalar case, and spectrum of matrix $Z_{,z}$ is stable in general case) and unstable part ($Z_{,z} > 0$ in scalar case, and at least one of eigenvalues of matrix $Z_{,z}$ has positive real part). In Fig. 3.1 parts AB and CD of Γ are stable, and part BD is unstable. System (3.16) is not stiff in a small neighborhood of the unstable part of Γ because the eigenvalues of matrix $f_{,x}$ have large positive real parts here.

The most interesting phenomenon takes place near the points B or D, where Γ loses stability and one of the eigenvalues of the matrix Z_z moves from the left part of the complex plane to the right half - plane, and the point $x(t)$ moves from one stable part of Γ into point C on another stable part of Γ. This happens within a short interval of time $O(\frac{\ln L}{L})$. This part of trajectory forms the so - called interior layer. Further, $x(t)$ moves according to the sign of Y along the surface Γ with the speed $O(1)$ to

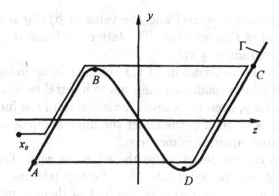

Figure 3.1.

the point D and so on. The same trajectory $x(t)$ is shown at Fig. 3.2.
Intervals of slow motion are interrupted by short $O(\frac{1}{L})$ boundary and
interior layers and it seems like the trajectory is discontinuous. Such a
situation is observed for trajectories of equations, describing processes
in chemical kinetics (see, for example, the trajectory behavior for the
problem of oregonator [39, 51]).

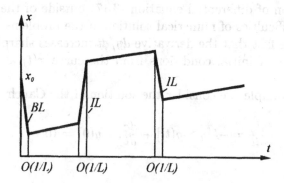

Figure 3.2.

Analysis of the simplest singular perturbed equation of form [95]

$$\varepsilon \frac{dy}{dt} = f(y, t), \qquad \varepsilon > 0. \tag{3.17}$$

leads to the same results.

Consider the case when a degenerated equation, corresponding to the
equation (3.17),

$$f(x, t) = 0$$

has a unique solution $x = x(t)$ and the value of $\partial f / \partial y$ is negative in the neighborhood of this solution. The latter condition is essential for stability of the solution $x = x(t)$.

The behavior of the equation (3.17) solution is as follows. If the value of ε is sufficiently small, the tangents to integral curves are almost parallel to the axis y, even for a small deviation from the function $x(t)$. And the smaller is value of ε, the faster the integral curve and solution $x(t)$ of degenerated equation come close.

The situation can be described in the following way. Two intervals of essentially different behavior take place for any integral curve in the domain considered. The duration of the first of them is much smaller than that of the second one. The first interval, where the desired function varies fast, represents the tending of the integral curve to the curve $x(t)$ and is named the boundary layer. In the second interval the derivatives are essentially smaller and the integral curve practically coincide with the curve $x(t)$. The boundary layer always takes place, except for the case when initial condition is a root of degenerate equation, i.e., when $y_0 = x(t_0)$. The smaller is value of parameter ε, the stronger is the difference of the behavior on the two intervals.

Thus, the solution of the degenerate equation can be used for describing the solution of differential equation (3.17) outside of the boundary layer. The difficulties of numerical solution of the problems considered result from the fact that the derivative dy/dt increases sharply even for small deviation of initial conditions from the curve $x(t)$ at any of its point.

As a first example, we consider the solution of the Cauchy problem

$$\frac{dy}{dt} = -k[y - g(t)] + \frac{dg}{dt}, \quad y(0) = 10, \qquad (3.18)$$

where $k \gg 1$, $g(t) = 10 - (10 + t)e^{-t}$.

In [28] for $k = 200$ the problem is named "artificial test problem of Lapidus and Zeinfeld". It has the analytical solution

$$y(t) = g(t) + 10e^{-kt}. \qquad (3.19)$$

Here we consider a more stiff case when $k = 1000$. We use for numerical solution the prediction - correction method in which the explicit and implicit Euler formulas are implemented for prediction and correction respectively. This program PC1 is described in chapter 2. The solution was computed on the interval $t \in [0, 1]$ with accuracy 10^{-5}.

After using the λ - transformation the problem (3.18) takes the form

$$\frac{dy}{d\lambda} = \frac{k(g-y)+\dot{g}}{\sqrt{1+(k(g-y)+\dot{g})^2}}, \qquad \frac{dt}{d\lambda} = \frac{1}{\sqrt{1+(k(g-y)+\dot{g})^2}}, \qquad (3.20)$$

$$y(0) = 10, \qquad t(0) = 0.$$

The execution time for this problem was three times less than that of the problem (3.18). It is explained by the fact that in order to obtain the results which agree well with the exact solution (3.19) for small values of t, i.e. in the boundary layer, very small integration step 10^{-5} with respect to t has to be used for the problem (3.18). Whereas for transformed problem (3.20) the admissible initial step with respect to argument λ was equal to 0.1.

The second problem, taken from [28], is nonlinear Edsberg's problem

$$\frac{dy}{dt} = -2ky^2, \qquad y(0) = 10, \qquad (3.21)$$

It has the exact analytical solution

$$y(t) = \frac{10}{1+20kt}.$$

After λ - transformation the problem (3.21) takes the form

$$\frac{dy}{d\lambda} = \frac{-2ky^2}{\sqrt{1+4k^2y^4}}, \qquad \frac{dt}{d\lambda} = \frac{1}{\sqrt{1+4k^2y^4}},$$

$$y(0) = 10, \qquad t(0) = 0.$$

This problem was integrated by the program PC1 for $k = 10^3$ with accuracy 10^{-10} and initial integration step 0.1 on the interval $t \in [0,1]$. The execution time for this problem was two times less then for the problem (3.18).

As another illustration of this type problem, we consider a well - known Dahlquist's special problem

$$\varepsilon\frac{dy}{dt} = (1-t)y - y^2, \qquad y(0) = 0.5, \qquad \varepsilon = 10^{-6}. \qquad (3.22)$$

Although this nonlinear equation is the Bernoulli's equation, it does not have an analytical solution. However, outside of the boundary layer, which is very small for this problem, numerical solution can be compared with the solution of degenerated problem $y = 1 - t$.

We consider the solution of the problem (3.22) on the interval $t \in [0, 1]$ with accuracy 10^{-7}. After using of the λ - transformation, the problem (3.22) takes the form

$$\frac{dy}{d\lambda} = \frac{(1-t)y - y^2}{\sqrt{\varepsilon^2 + \left((1-t)y - y^2\right)^2}}, \quad \frac{dt}{d\lambda} = \frac{\varepsilon}{\sqrt{\varepsilon^2 + \left((1-t)y - y^2\right)^2}}$$

$$y(0) = 0.5, \qquad t(0) = 0.$$

$$(3.23)$$

The program PC1, described in chapter 2, required a lot of execution time for this problem. Therefore, an attempt was made to modify the simple iteration process in order to achieve reasonable execution time. The iterative process of solution of the nonlinear system of equation

$$y_{m+1} = y_m + hf(y_{m+1}, t_{m+1}), \qquad m = 0, 1, 2, \ldots,$$

that implements implicit Euler method for Cauchy problem

$$\frac{dy}{dt} = f(y, t), \qquad y(t_0) = y_0, \qquad y \in {}^n, \qquad f = (f_1, \ldots, f_n)^T$$

was constructed on the basis of Newton - Raphson method, in which only the elements on the main diagonal of Jacoby matrix were used. In this case, the iterative process is described by the formula

$$y_{i,m+1}^{(k+1)} = y_{i,m+1}^{(k)} - \frac{y_{i,m+1}^{(k)} - y_{i,m} - hf_i(y_{m+1}^{(k)}, t_{m+1})}{1 - h\dfrac{\partial f_i(y_m, t_m)}{\partial y_{i,m}}}$$

$$(3.24)$$

$$m = 0, 1, 2, \ldots; \qquad k = 0, 1, 2, \ldots; \qquad i = \overline{1, n}.$$

and the initial approximation of vector function y calculated on the $m+1$ - th integration step is determined by explicit Euler method.

The following algorithm was used: if iterative process (2.30) has not converged with a given accuracy after 10 iterations, the iteration process (3.24) is used. For this process diagonal elements of Jacoby matrix are not calculated at each iteration. If the latter process has not converged after 10 iterations, we start calculating these diagonal elements at each iteration. If this process has not converged either, the integration step is bisected.

This algorithm was coded in PC1M program. This program integrated the problem (3.23) in 5 dimensionless minutes.

Programs PC1 and PC1M implemented one-step method of the first order of accuracy. Such methods are often used in large computations of reacting currents in two and three dimensional problems [88], because they need to minimize the information stored in computer memory. However, multi - step methods have higher order of accuracy and converge faster. Therefore, three more programs PC2, PC3, PC4 were developed. These programs are based on multi - step back differentiation formulas [42] of second, third, and fourth order of accuracy respectively.

In [7, 95] it is recommended to calculate the prediction by extrapolation formulas which use only the solution vector in preceding points.

The program PC2 calculates the first prediction as

$$y_{m+1}^p = 2y_m - y_{m-1},$$

and the correction is found by the formula

$$y_{m+1}^c = \frac{1}{3}(4y_m - y_{m-1} + 2hf(y_{m+1}^p, t_{m+1})). \qquad (3.25)$$

If the step is doubled, the previously calculated results are used. If the step is bisected, the solution at point $m - 1/2$ is calculated by the formula

$$y_{m-\frac{1}{2}} = \frac{y_{m-1} + y_m}{2}$$

The program PC3 calculates the first prediction as

$$y_{m+1}^p = y_{m-2} - 3y_{m-1} + 3y_m,$$

and correction is found by the formula

$$y_{m+1}^c = \frac{18y_m - 9y_{m-1} + 2y_{m-2} + 6hf(y_{m+1}^p, t_{m+1})}{11}. \qquad (3.26)$$

The missing values of y for bisected step are calculated as

$$y_{m-\frac{1}{2}} = \frac{3y_m + 6y_{m-1} - y_{m-2}}{8}.$$

When step is doubled they are determined as

$$y_{m-3} = y_m - 3y_{m-1} + 3y_{m-2}.$$

The program PC4 calculates first prediction by the formula

$$y_{m+1}^p = 4y_m - 6y_{m-1} + 4y_{m-2} - 4y_{m-3}, \qquad (3.27)$$

and correction is found as

$$y^c_{m+1} = \frac{48y_m - 36y_{m-1} + 16y_{m-2} - 3y_{m-3} + 12hf(y^p_{m+1}, t_{m+1})}{25}.$$

(3.28)

Missing values for bisected step are counted in according to the expression

$$y_{m-\frac{1}{2}} = \frac{5y_m + 15y_{m-1} - 5y_{m-2} + y_{m-3}}{16},$$

(3.29)

$$y_{m-\frac{3}{2}} = \frac{-y_m + 9y_{m-1} + 9y_{m-2} - y_{m-3}}{16}.$$

(3.30)

When the step is doubled, a new value of y_{m-2} is not calculated. It is stored in memory. The value of y_{m-3} is calculated by the formula

$$y_{m-3} = -5y_m + 16y_{m-\frac{1}{2}} - 15y_{m-1} + 5y_{m-2}.$$

(3.31)

As an example, let us consider how these formulas were obtained for program PC4 in more detail. The formula of correction (3.28) was taken from [42]. Extrapolation formula of prediction (3.27) was obtained as a solution of the following system of equations

$$
\begin{aligned}
y_{m+\alpha} &= y_m + \alpha hy' + \frac{(\alpha h)^2}{2!}y'' + \frac{(\alpha h)^3}{3!}y''' + O(h^4), \\
y_{m-1} &= y_m - hy' + \frac{h^2}{2!}y'' - \frac{h^3}{3!}y''' + O(h^4), \\
y_{m-2} &= y_m - 2hy' + 2h^2y'' - \frac{4}{3}h^3y''' + O(h^4), \\
y_{m-3} &= y_m - 3hy' + \frac{9}{2}h^2y'' - \frac{9}{2}h^3y''' + O(h^4).
\end{aligned}
$$

(3.32)

Here the derivatives are calculated in point y_m and the value of the parameter α is equal to $\alpha = 1$.

Formulas (3.29), (3.30) are obtained if the parameter is equal to $\alpha = -0.5$ and $\alpha = -1.5$ respectively.

Finally, the formula (3.31) can be obtained by solving relation (3.29) with respect to y_{m-3}.

The starting routine for the programs PC2, PC3, PC4 is the modified Euler method.

In addition to these programs, the programs PC2M, PC3M, PC4M based on the same principle as the program PC1M described above were also developed. I.e., an iterative process based on the Newton -Raphson method, in which only elements of main diagonal of Jacoby matrix are taken in account, was implemented.

The iteration process in program PC2M is designed on the basis of the formula (3.25) in the form

$$y_{i,m+1}^{(k+1)} = y_{i,m+1}^{(k)} - \frac{y_{i,m+1}^{(k)} - \frac{1}{3}(4y_{i,m} - y_{i,m-1} + 2hf_i(y_{m+1}^{(k)}, t_{m+1}))}{1 - \frac{2}{3}h\frac{\partial f_i(y_m, t_m)}{\partial y_{i,m}}},$$

$$m = 0, 1, 2, \ldots, \qquad k = 0, 1, 2, \ldots, \qquad i = \overline{1, n}.$$

The iteration process in program PC3M uses the formula (3.26) and takes the form

$$y_{i,m+1}^{(k+1)} = y_{i,m+1}^{(k)} - \frac{y_{i,m+1}^{(k)} - \frac{1}{11}(18y_{i,m} - 9y_{i,m-1} + 2y_{i,m-2} + 6hf_i(y_{m+1}^{(k)}, t_{m+1}))}{1 - \frac{6}{11}h\frac{\partial f_i(y_m, t_m)}{\partial y_{i,m}}},$$

$$m = 0, 1, 2, \ldots, \qquad k = 0, 1, 2, \ldots, \qquad i = \overline{1, n}.$$

Finally, the iteration process in the program PC4M is described in accordance with the formula (3.28) by the expression

$$y_{i,m+1}^{(k+1)} =$$
$$= y_{i,m+1}^{(k)} - \frac{y_{i,m+1}^{(k)} - \frac{1}{25}(48y_{i,m} - 36y_{i,m-1} + 16y_{i,m-2} - 3y_{i,m-3} + 12hf_i(y_{m+1}^{(k)}, t_{m+1}))}{1 - \frac{12}{25}h\frac{\partial f_i(y_m, t_m)}{\partial y_{i,m}}},$$

$$m = 0, 1, 2, \ldots, \qquad k = 0, 1, 2, \ldots, \qquad i = \overline{1, n}.$$

Problem (3.23) was solved by the program PC2M in 2.3 dimensionless minutes. The solution of this problem by program PC3M takes 6 dimensionless minutes of execution time. Apparently, the linear prediction used in the programs PC1M and PC2M is more efficient for finding the solution outside of boundary layer. It was shown above that the linear function $x = 1 - t$ is a good approximates to the solution in this region. This function is a solution of corresponding degenerated equation. This is also confirmed by the fact that numerical results agree with this function $x = 1 - t$ with high accuracy.

As another example, we consider numerical solution of Cauchy problem for Van der Pol's equation [53, 51]. If a new variable $t = x/\mu$ is introduced in classical Van der Pol's equation $y'' - \mu(1 - y^2)y' + y = 0$, the equation take the form $\varepsilon y'' - (1 - y^2)y' + y = 0$ where $\varepsilon = 1/\mu^2$. And

Cauchy problem can be formulated in the form

$$\frac{dy_1}{dt} = y_2, \quad \varepsilon\frac{dy_2}{dt} = (1 - y_1^2)y_2 - y_1,$$

$$y_1(0) = 2, \quad y_2(0) = -0.66$$

Here the small parameter is $\varepsilon = 10^{-6}$.

This is a well - known benchmark for estimation the efficiency of computational programs intended to solve stiff systems.

The program PC1 was used for solving this problem. The accuracy of computations was checked by comparison with "exact results" obtained in [53] and it did not exceed the value of 10^{-3}. The problem was solved for $t \in [0, 0.01]$. The λ - transformation reduced execution time tenfold. The number of calculations of the right - hand sides was reduced twenty fold. The integration step of the λ - transformed problem was ten times bigger.

Note, that the program PC1 allows to find the solution of λ - transformed Van der Pol problem for $t \in [0, 2]$ whereas the solution the original problem terminated at $t = 0.03$ because of exponent overflow.

3. Stiff systems

Starting consideration of stiff system of ordinary differential equations it is should be noted that they can be considered as systems which can be obtained from singular - perturbed systems by means of some unknown change of variables. Thus, stiff systems are systems with numerous implicit parameters because there is no explicit group of variables that can be divided into fast and slow ones (as z and y for system (3.16)) and simple equation of surface Γ similar to $Z(y, z) = 0$ for the system (3.16) is not known. Therefore, the asymptotic theory, similar the theory for singular - perturbed system, does not exist.

An attempt was made in [34, 35, 33] to develop such a theory for a special type of stiff systems called regular ones. In [34, 35] this theory was implemented for analysis of some numerical methods of solution of stiff systems.

Let us consider the solution of some stiff problems using the λ - transformation.

As a first example we investigate the longitudinal perturbed motion of an airplane [95]. If we consider linear stable flight without sliding and with small deviations of parameters from their initial values, the equations of perturbed motion are divided into two independent systems describing longitudinal and lateral motions. Let us use velocity coordinate axis. The origin is in center of gravity of the airplane, x axis is along the flight velocity v, y axis is orthogonal to x axis and it is on airplane symmetry plane and z axis is directed along the right airplane wing. Then the following system of equations takes place

$$m\frac{dv}{dt} = P - Q - G\sin\theta, \quad mv\frac{d\theta}{dt} = Y - G\cos\theta,$$

$$I_z\frac{d^2\vartheta}{dt^2} = M_z - Py_p, \quad \alpha = \vartheta - \theta, \tag{3.33}$$

where m is the mass of airplane; y_p is the arm of the engine thrust force with respect to the center of gravity of the airplane; v is the velocity; P is the force of thrust acting along the engine axis; Y is the lift force directed upwards and orthogonal to the flight velocity; G is the weight of airplane directed downwards vertically; Q is the head resistance directed along the velocity; M_z is the moment of external forces with respect to OZ axis; I_z is the moment of inertia of the airplane with respect to OZ axis; θ is the angle between vector of speed and horizontal plane; ϑ is the pitch angle; α is the angle of attack.

In general case, forces and moments entering in system (3.33) depend on many parameters of motion: the angle of attack α, the air density ρ, the flight velocity, the deviation angle of elevator, the pitch angle and their derivatives with respect to time.

Basing on the method of small perturbations and using the dimensionless form we obtain the system of equations for perturbations

$$\Delta v = \frac{v - v_0}{v_0}, \quad \Delta\alpha = \alpha - \alpha_0, \quad \Delta\vartheta = \vartheta - \vartheta_0, \quad \tau = \frac{\rho s v_0}{2m}t,$$

where s is the wing area. Everywhere below the symbol Δ standing before v, α and ϑ will be omitted.

For the horizontal flight and specific values of airplane parameters system of equations (3.33) takes the form

$$\frac{dv}{d\tau} = -0.104v + 0.043\alpha - 0.1\vartheta, \quad \frac{d\alpha}{d\tau} = -0.57v - 5.12\alpha + \omega_z,$$

$$\frac{d\vartheta}{d\tau} = \omega_z, \quad \frac{d\omega_z}{d\tau} = -12.574v - 43.69\alpha - 9.672\omega_z. \tag{3.34}$$

Roots of the characteristic equation of this system, given in [95], are equal to $\lambda_{1,2} \approx -7.4 \pm 6.2i$, $\lambda_3 \approx -0.27$, $\lambda_4 \approx 0.16$, $i^2 = -1$ and they are divided on two groups with respect their modulus $|\lambda_1| = |\lambda_2| \gg |\lambda_3| > |\lambda_4|$,

This is typical for horizontal motion of airplane and is stipulated by physics of process considered. The motion which corresponds to the roots with large modulus is called short - period one. The motion corresponding to two smaller roots is called long - period one.

Thus, system (3.34) is stiff for any interval $[0,\ T]$ the duration of which exceeds considerably the duration of boundary layer ($\tau_{bs} < 0.1$).

The system remains stiff for climb and glide conditions, although difference between roots decreases.

The system describing lateral motion of an airplane, as well as the general system unifying longitudinal and lateral motions, are also stiff.

Note, that similar difference of motions is typical not only for airplane but for rocket as well.

To estimate an accuracy of the numerical solution, we obtain the analytical solution of system of linear differential equations (3.17)

$$v = ae^{\lambda t}, \qquad \alpha = be^{\lambda t}, \qquad \vartheta = ce^{\lambda t}, \qquad \omega_z = de^{\lambda t}. \qquad (3.35)$$

It is accepted here that $t = \tau$. Substituting (3.35) in (3.34) we obtain the system of linear algebraic equations for $a,\ b,\ c,\ d$

$$
\begin{aligned}
-(0.104 + \lambda)a\ +\quad &0.043b\ -\ 0.1c &&= 0, \\
-0.57a\ -\ &(5.12 + \lambda)b && +\qquad\quad d = 0, \\
&- \lambda c\ + &&\qquad\quad d = 0, \\
-12.574a\ -\ &43.69b && -\ (9.672 + \lambda)d = 0,
\end{aligned}
$$
$$(3.36)$$

This system is homogeneous. For the system to have nontrivial solution its determinant should be equal to zero

$$
\begin{vmatrix}
-0.104 - \lambda & 0.043 & -0.1 & 0 \\
-0.57 & -5.12 - \lambda & 0 & 1 \\
0 & 0 & -\lambda & 1 \\
-12.574 & -43.69 & 0 & -9.672 - \lambda
\end{vmatrix} = 0.
$$

Calculating the determinant, we obtain the characteristic equation

$$\lambda^4 + 14.896\lambda^3 + 94.774\lambda^2 + 9.215\lambda - 3.948 = 0,$$

Its roots calculated by means of Newton - Raphson method take the values

$$\lambda_1 = 0.1596, \qquad \lambda_2 = -0.2651, \qquad \lambda_{3,4} = -7.3952 \pm i6.211. \quad (3.37)$$

Taking into account system (3.36) and assuming $c = 1$, we obtain that eigenvalues (3.37) correspond to the following values of parameters

a, b, c, d

$$\lambda_1: \quad a = -0.36795, \quad b = 0.06996, \quad c = 1, \quad d = 0.1596,$$
$$\lambda_2: \quad a = 0.6563, \quad b = -0.1316, \quad c = 1, \quad d = -0.265,$$
$$\lambda_3 = -7.3952 - 6.211i: \quad a = 0.57385 \cdot 10^{-2} - i0.5971 \cdot 10^{-3},$$
$$b = 1.2664 - i0.7274, \quad c = 1, \quad d = -7.3952 - i6.211.$$

Therefore, the general solution of system (3.34) takes the form

$$\begin{pmatrix} v \\ \alpha \\ \vartheta \\ \omega_z \end{pmatrix} = C_1 \begin{pmatrix} -0.36795 \\ 0.06996 \\ 1 \\ 0.1596 \end{pmatrix} e^{0.1596t} + C_2 \begin{pmatrix} 0.6563 \\ -0.1316 \\ 1 \\ -0.265 \end{pmatrix} e^{-0.265t} +$$

$$+ \left(\cos 6.21t \cdot \left[C_3 \begin{pmatrix} 0.57385 \cdot 10^{-2} \\ 1.2664 \\ 1 \\ -7.395 \end{pmatrix} + C_4 \begin{pmatrix} -0.5971 \cdot 10^{-3} \\ -0.7274 \\ 0 \\ -6.211 \end{pmatrix} \right] - \right.$$

$$\left. - \sin 6.21t \cdot \left[C_3 \begin{pmatrix} 0.5971 \cdot 10^{-3} \\ 0.7274 \\ 0 \\ 6.211 \end{pmatrix} + C_4 \begin{pmatrix} 0.57385 \cdot 10^{-2} \\ 1.2664 \\ 1 \\ -7.395 \end{pmatrix} \right] \right) e^{-7.395t},$$

where C_1, C_2, C_3, C_4 are arbitrary constants determined from initial conditions. If initial conditions are taken in the form

$$t = 0, \qquad v = \vartheta = \omega_z = 0, \qquad \alpha = 1, \qquad (3.38)$$

then

$$C_1 = -0.2969, \qquad C_2 = -0.17106,$$
$$C_3 = 0.46798, \qquad C_4 = -0.5576.$$

The problem (3.34),(3.38) was integrated on the interval $t \in [0, 5]$ with the program PC1. For the same accuracy of computations the execution time after λ - transformation was two times less than without the transformation.

We consider now the stiff system of ordinary differential equations arising in chemical kinetics [39]

$$\frac{dy_1}{dt} = 77.27(y_2 + y_1(1 - 8.375 \cdot 10^{-6}y_1 - y_2))$$

$$\frac{dy_2}{dt} = \frac{1}{77.27}(y_3 - (1 + y_1)y_2) \qquad (3.39)$$

$$\frac{dy_3}{dt} = 0.161(y_1 - y_3)$$

This system describes a famous chemical reaction with limiting cycle, the "oregonator". It is the reaction of $HBrO_2$ with Br^- and $Ce(IV)$. System of equations (3.39) was integrated in [39] for initial conditions

$$y_1(0) = 3, \qquad y_2(0) = 1, \qquad y_3(0) = 2. \qquad (3.40)$$

Solving of the problem by ordinary prediction - correction method failed because of high stiffness of the problem. The solution was obtained only by method given in [42].

The diagrams of solution of problem (3.39), (3.40) (see Fig. 3.3, 3.4) are given in [51]. It is noted there that the system is an example of a stiff one. Its solution changes by many orders of magnitude. Therefore, the given example can be used as a severe test for the programs of numerical integration.

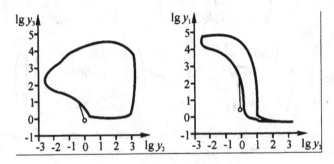

Figure 3.3.

An attempt to solve the problem (3.39), (3.40) by means of the program PC1 was not successful. But after using the λ - transformation, the solution was obtained for the whole cycle, i.e., for $t \in [0, 303]$. The dimensionless execution time was very large (about one hour). After applying the λ - transformation to the problem (3.39), (3.40) it takes the form

$$\frac{dy_1}{d\lambda} = \frac{P_1}{Q}, \qquad \frac{dy_2}{d\lambda} = \frac{P_2}{Q}, \qquad \frac{dy_3}{d\lambda} = \frac{P_3}{Q}, \qquad \frac{dt}{d\lambda} = \frac{1}{Q}, \qquad (3.41)$$

$$y_1(0) = 3, \qquad y_2(0) = 1, \qquad y_3(0) = 2, \qquad t(0) = 0.$$

where P_i are the right - hand sides of equations (3.39),

$$Q = \sqrt{1 + P_1^2 + P_2^2 + P_3^2}.$$

Several attempts were made to decrease the execution time.

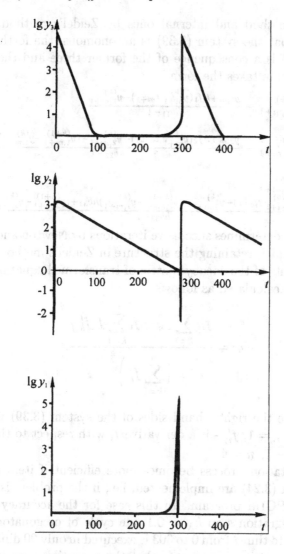

Figure 3.4.

An attempt to implement an iterative process for $(m+1)$ - th step by means of the formula

$$y_{i,m+1}^{(k+1)} = y_{i,m+1}^{(k)} - \Omega_i \left[y_{i,m} + h f_i(y_{m+1}^{(k)}, t_{m+1}) - y_{i,m+1}^{(k)} \right],$$

in which parameters Ω_i was selected taking into account the norm of the Jacoby matrix of system that provide a convergence conditions was not successful.

Situation improved substantially when the iteration scheme proposed in [102] was implemented. In this scheme external iterations are made by Newton's method and internal ones by Zeidel's method. Taking into account that the system (3.39) is autonomous, the forth equation of system(3.41) is a consequence of the former three and the iterative scheme in the case takes the form

$$y_{1,m+1}^{(k+1)} = y_{1,m+1}^{(k)} - \frac{y_{1,m}+hf_1(y_{m+1}^{(k)},t_{m+1})-y_{1,m+1}^{(k)}}{hb_{11}-1},$$

$$y_{2,m+1}^{(k+1)} = y_{2,m+1}^{(k)} - \frac{y_{2,m}+hf_2(y_{m+1}^{(k)},t_{m+1})-y_{2,m+1}^{(k)}-(y_{1,m+1}^{(k+1)}-y_{1,m+1}^{(k)})hb_{21}}{hb_{22}-1},$$

$$y_{3,m+1}^{(k+1)} = y_{3,m+1}^{(k)} -$$
$$- \frac{y_{3,m}+hf_3(y_{m+1}^{(k)},t_{m+1})-y_{3,m+1}^{(k)}-(y_{1,m+1}^{(k+1)}-y_{1,m+1}^{(k)})hb_{31}+(y_{2,m+1}^{(k+1)}-y_{2,m+1}^{(k)})hb_{32}}{hb_{33}-1}.$$

This scheme determines successive iterations by Newton and only one iteration by Zeidel, retaining the structure of Zeidel's method. Here b_{ij} - are components of the Jacoby matrix of transformed equations system (3.41) that are calculated as follows

$$b_{ij} = \frac{f'_{ij}\sum_{k=1}^{n} f_k^2 - f_i \sum_{k=1}^{n} f_k f'_{k,j}}{\left(\sum_{k=1}^{n} f_k^2\right)^{\frac{3}{2}}}. \tag{3.42}$$

Here f_i - are the right - hand sides of the system (3.39) where it is assumed that $f_n = 1$, f'_{ij} - is a derivative f_i with respect to the variable y_j, i, $j = \overline{1, n-1}$, $n = 4$.

The computation process becomes more efficient if iterations given by the formula (3.24) are implemented, i.e., if the problem is solved by means of the PC1M program. In this case for the accuracy 10^{-4} and the initial integration step $h_\lambda = 0.1$ one cycle of oregonator behavior corresponding to time t from 0 to 303 is executed in only 90 dimensionless seconds . It is interesting to note that the execution time increases if simple iteration is excluded from the program. I.e., the best computation process is provided by using the iterative formulas (2.30) and (3.24) together.

A program, which takes into account not only diagonal element of Jacoby matrix but also one or two offdiagonal elements with maximal absolute values, was developed . For example, if maximal offdiagonal element is b_{ij} then i - th formula of the iterative scheme (3.24) should be corrected by the term

$$\frac{(y_{j,m} + hf_j(y_{m+1}^{(k)},t_{m+1}) - y_{j,m+1}^{(k)})hb_{ij}}{(hb_{ii} - 1)(hb_{jj} - 1)}$$

Computations show that such a correction does not reduce execution time. However, the minimal integration step is increased. Apparently, it is explained by the fact that this increase of the step is compensated by complexity of the algorithm.

For solving the untransformed problem (3.39), (3.40) the program DLSODE was used. This program is one of the latest versions of the Geer program for integration of stiff systems of equations. This program was taken from the package ODEPACK assembled by Hindmarsh [57]. The execution time was equal to 13 seconds.

The problem (3.41) was solved in 70, 20, and 17 dimensionless seconds by means of the PC2M, PC3M, and PC4M programs correspondently.

The following conclusions can be made:

— the program PC1M requires more execution time than program DLSODE but it does not require the storage of information obtained at the previous steps. This can be a decisive factor for solving multy - dimensional problems of solid mechanics which reduce to systems of ordinary differential equations of high order.

The execution time of the program PC4M is about the same as that of the program DLSODE. But it certainly has an advantage. Namely, it can be used for integration of problems which solutions are closed curves, which means that their right - hand sides are infinite at some points.

We consider now solution of the model stiff problem proposed by I. Babushka [16, 74, 75]. Taking into account that one of integral of this problem is known we can represent the problem in the form

$$\frac{dy_1}{dt} = a\frac{y_1 y_3}{y_2}, \qquad \frac{dy_2}{dt} = -ay_3,$$

$$\frac{dy_3}{dt} = \left(cy_4 - 9a\frac{y_1 y_3}{y_2} + (c + ay_3)y_3 - b\right) \cdot \frac{1}{y_2 + 9},$$

$$\frac{dy_4}{dt} = a\frac{y_1 y_3}{y_2} - d(y_3 + y_4),$$

$$y_1(0) = y_2(0) = 1, \qquad y_3(0) = 0, \qquad y_4(0) = 0.1, \qquad t \in [0, 1],$$

$$a = d = 100, \qquad b = 0.9, \qquad c = 1000.$$

It is noted in [74] that the Jacoby matrix of this system has both positive and negative values sufficiently far from zero. That is why the integration step for explicit and implicit methods does not increase and remains very small with the increase of time. It also causes additional difficulties in iteration method.

After using the λ - transformation this problem was solved by the program PC3M in 24 seconds for PC AT 286/287. Maximal error for $t = 1$ was equal to 0.04. Note, that such an accuracy can not be obtained for untransformed problem.

The transformed problem was solved also by PC486. To obtain the same accuracy the program DUMKA [74, 75] took 1.4 seconds and program DLSODE [57] takes 0.05 seconds. Right - hand sides were calculated 36 074 times, 4.225 times, and 520 times for the programs DUMKA, PC3M, and DLSODE respectively.

The "exact" solution was obtained by means of the program RADAU5, DLSODE and for $t = 1$ it was equal to

$$y_1 = 0.58367615856 \cdot 10^5, \qquad y_2 = 0.171327881975 \cdot 10^{-4},$$

$$y_3 = 0.1903613486 \cdot 10^{-5}, \qquad y_4 = 0.583667158388 \cdot 10^4.$$

4.   Stiff equations for partial derivatives

The λ - transformation was used above for the problems of small dimensionality ($n \leq 4$). Now, we consider solution of stiff system of ordinary differential equations of high dimensionality ($n > 10$) to which the solution of equations for partial derivations is reduced [68].

For example, we consider stiff problem describing a moving rectangular plate when a car runs over it [53]. Deflections of the plate $W(x, y, t)$ are described by the following equations with boundary and initial conditions

$$D\nabla^2\nabla^2 W + G\frac{\partial W}{\partial t} + \frac{\partial^2 W}{\partial t^2} = f(x, y, t),$$

$$W|_{t=0} = 0, \quad \frac{\partial W}{\partial t}\Big|_{t=0} = 0, \quad W|_{(x,y)\in\partial\Omega} = 0, \quad \nabla^2 W|_{(x,y)\in\partial\Omega} = 0.$$

Here $\nabla^2\nabla^2$ - is biharmonic operator; D=100, G=1000 - are parameters of the plate rigidity and friction; ∇^2 - is Laplace operator; $\partial\Omega$ - is boundary of domain $\Omega = \{(x, y), \quad 0 \leq x \leq 2, \quad 0 \leq y \leq 4/3\}$ which is divided on meshes by points $x_i = ih, y_j = jh, h = 2/9, i = \overline{1,8}, j = \overline{1,5}$.

The load $f(x, y, t)$ is approximated by the sum of two Hessians that describe the motion of four wheels along x axis

$$f(x, y, t) = \begin{cases} 200(e^{-5(t-x-2)^2} + e^{-5(t-x-5)^2}), & y = y_1, y = y_4, \\ 0, & y \neq y_1, y \neq y_4. \end{cases}$$

Biharmonic operator is represented in standard finite - difference form using central differences

$$\nabla^2\nabla^2 W_{i,j}(t) = (W_{i-2,j} - 8W_{i-1,j} + 20W_{i,j} - 8W_{i+1,j} +$$

$$+W_{i+2,j} + W_{i,j-2} - 8W_{i,j-1} - 8W_{i,j+1} + W_{i,j+2} + 2W_{i+1,j-1} +$$

$$+2W_{i+1,j+1} + 2W_{i-1,j-1} + 2W_{i-1,j+1})/h^4.$$

In accordance with the boundary conditions the deflections of plate $W_{i,j}(t) = 0$ is equal to zero for the boundary points and it is two times less for pre boundary points comparably to the points that on normal to the contour near the boundary.

The obtained system of ordinary differential equations was integrated for $[0, t], t \in [0, 7]$ by the program PC1. Because of small absolute values of deflection, which was no more than 10^{-2}, we checked the relative error rather than absolute error. The relative error was obtained by dividing the absolute error by the sum of square values of $W_{i,j}$.

It was found that application of λ - transformation reduced the execution time thirtyfold. The number of required right hand sides computations reduced sixteen times. The integration step increased more than ten times.

Chapter 4

DIFFERENTIAL – ALGEBRAIC EQUATIONS

Differential - algebraic equations (DAE) differ from other problems with solutions given by smooth and continuous parametric sets. They combine specifics of the nonlinear algebraic or transcendental equations with those of ODE. Correct formulation of the Cauchy problem for such equations requires solution of a system of nonlinear equations. In this chapter we will describe and investigate the algorithm of numerical continuation of Cauchy problem solution for different forms of DAE.

1. Classification of systems of DAE

We remind that the following system is named a system of implicit ordinary differential equations

$$\Phi(y, \dot{y}, t) = 0,$$

$$y(t) : {}^1 \to {}^n, \quad t \in [t_0, T], \quad \Phi : {}^{2n+1} \to {}^n, \quad \dot{y} = \frac{dy}{dt}. \tag{4.1}$$

If the matrix $\partial\Phi/\partial\dot{y}$ is nonsingular the system can be resolved with respect to the derivatives and represented in explicit form (i.e. normal Cauchy form)

$$\dot{y} = \varphi(y, t); \qquad f = (\varphi_1, \ldots, \varphi_n)^T. \tag{4.2}$$

A generic form of a system of differential - algebraic equations is as follows

$$F(y, \dot{y}, x, t) = 0,$$

$$G(y, x, t) = 0, \tag{4.3}$$

$$y(t) : {}^1 \to {}^n, \quad x(t) : {}^1 \to {}^m, \quad F : {}^{2n+m+1} \to {}^n,$$

$$G : {}^{n+m+1} \to {}^m, \quad t \in [t_0, T].$$

It consists of ordinary differential equations F and nondifferential relations G. The latter are usually nonlinear algebraic or transcendental equations.

It is assumed for system (4.3) that the matrix $\partial F/\partial \dot{y}$ is nonsingular. It is obvious that system (4.1) can be reduced to a system of this type if matrix $\partial \Phi/\partial \dot{y}$ of (4.1) is singular. Thus, the term "singular system of equations" is often used instead of more traditional term "differential - algebraic equations" (see, as example, [17, 18, 11]) to denote the system (4.3) .

Close connection exists between differential - algebraic and singular perturbed equations. It has been shown above that the latter belong to the group of stiff systems of differential equations. Indeed, when a small parameter ε is present in a singular perturbed equations they become stiff. If the parameter ε is turned to zero they are transformed in differential - algebraic equations.

We consider the Van - der - Pol equation that was investigated in the classical paper [27]. We write it in the form

$$\varepsilon z'' + (z^2 - 1)z' + z = 0.$$

Since

$$\varepsilon z'' + (z^2 - 1)z' = \frac{d}{dt}\left(\varepsilon z' + \left(\frac{z^3}{3} - z\right)\right) = -z,$$

then, if notation $y = \varepsilon z' + \frac{z^3}{3} - z$ is introduced, we have a system

$$y' = -z,$$

$$\varepsilon z' = y - \left(\frac{z^3}{3} - z\right),$$

which, in general case, can be written as

$$y' = f(y, z),$$

$$\varepsilon z' = g(y, z).$$

This system of singular perturbed equations will be stiff for a small parameter ε. If this parameter ε is turned to zero, we obtain the system of differential - algebraic equations

$$y' = -z,$$

$$y - \left(\frac{z^3}{3} - z\right) = 0, \tag{4.4}$$

or, in general case,

$$y' = f(y, z),$$

$$g(y, z) = 0.$$

The problem (4.4) has exact solution

$$\ln|z| - \frac{z^2}{2} = t + C.$$

This solution is realized if the initial values are found on the curve L that is determined by the second equation of the system (4.4). The points of the curve with coordinates $y = \pm 2/3$, $z = \mp 1$ in which derivation $g_z = \partial g/\partial z$ becomes zero are of particular interest. The branch $-1 < z < 1$ of the curve L is unstable ($g_z > 0$), thus, the solution jumps onto the other stable branch ($g_z < 0$) at these points.

The systems of differential - algebraic equations are characterized by differential index. At first, we consider the index for the system of the form (4.1). This system has differential index r if r is minimal number of differentiations

$$\Phi(y, \dot{y}, t) = 0, \quad \frac{d\Phi(y, \dot{y}, t)}{dt} = 0, \ldots, \quad \frac{d^r \Phi(y, \dot{y}, t)}{dt^r} = 0,$$

transforming (4.1) to the explicit system (4.2).

It follows from the definition that if the matrix of system (4.1) $\partial \Phi/\partial \dot{y}$ is not singular, the differential index of this system is equal to zero. In this case the system reduces to canonical form (4.2) without differentiation.

Differential index of system (4.3) is determined in the following way. Since the matrix $\partial F/\partial \dot{y}$ is nonsingular this system can be represented in the form

$$\dot{y} = f(y, x, t),$$

$$G(y, x, t) = 0. \tag{4.5}$$

The index of the system is equal to one if after differentiation of its second equation with respect to t we obtain the following system

$$G_{,y}\,\dot{y} + G_{,x}\,\dot{x} + G_{,t} = 0, \tag{4.6}$$

where $G_{,x}$ is nonsingular. Thus, the system (4.5) can be written in the normal Cauchy form

$$\dot{y} = f(x, y, t)$$

$$\dot{x} = -G_{,x}^{-1}\left(G_{,y}\, f + G_{,t}\right).$$

If matrix $G_{,x}$ is singular the expression (4.6) takes the form

$$G_{,y}\, f + G_{,t} = 0$$

and it must be differentiated one more time. Differentiation continues until the matrix in front of the derivative $\dot{x}$ becomes nonsingular.

Thus, the index of a system of differential - algebraic equations (4.3) is equal to the minimal number of differentiations with respect to t that allows to determine $\dot{x}$ as a continuous function of y, x, t.

Note that system of equations (4.1) with nonsingular matrix $\partial\Phi/\partial\dot{y}$ and index equal to zero can be reduced to the system of the type (4.5) by introducing a new function $\dot{y} = x$

$$\dot{y} = x,$$

$$\Phi(y, x, t) = 0.$$

(4.7)

Now the index of this system is equal to one, i.e., the transformation (4.1) to the form (4.7) increases the differential index by one [13].

Here we consider a numerical solution of the Cauchy problem for a system of differential - algebraic equations with index equal to zero or one. It is assumed, nevertheless, that the derivative $G_{,x}$ can be singular in certain points. Many practical numerical problems are represented by such systems.

Solution of differential - algebraic equations is a more complicated problem comparably to the solution of ordinary differential equations. The following difficulties were noted in [13]:

1) initial conditions must be consistent with nondifferentiated relations;

2) the system of linear algebraic equations solved at each step of integration is ill - conditioned for a short step. It is shown in [13] that the order of conditionality of the system is $O(h^\nu)$ where ν is the index of the system and h is the integration step;

3) an error associated with the choice of step of integration is sensitive to the inconsistency of initial conditions and to a sharp change of the solution;

4) numerical solution is more sensitive to the accuracy of the approximation of the iterative matrix comparably to the solution of a system of ordinary differential equations.

The approach suggested here reduces some of these difficulties.

Concerning the consistency, the initial conditions for the system (4.3) can be taken in form

$$y(t_0) = y_0, \qquad x(t_0) = x_0,$$

where the vectors y_0 and x_0 must comply with the system of equations

$$G(y_0, x_0, t_0) = 0.$$

To our knowledge the numerical solution of differential - algebraic equations was examined first in [43]. In this work a system of equations describing processes in electric circuits was integrated by means of the backward differentiation formulas. The problem was to solve a system of linear with respect to the derivatives $\dot{y}$ equations

$$A(y)\dot{y} + B(y) = f(t), \qquad (4.8)$$

where $A(y)$ is matrix.

By present several monographs have been published [53, 17, 18, 11, 13, 47, 52] in addition to numerous papers on this subjects. In [56] the program LSODI intended for the integration of the Cauchy problem for implicit systems of ordinary differential equations was described. A program for solving the Cauchy problem for system of the form (4.8) which appear in electrical engineering was presented in [112]. Both programs use a version of the program GEAR which is based on backward differential formulas. The program DASSL [13], [93] was designed for solving initial value problem for an implicit system of the form (4.1) with index equal zero or one. In accordance with general approach suggested in [43], derivative is replaced by finite difference approximation that uses the backward differentiation formulas. The system of nonlinear equations obtained is solved by Newton's method in which the Jacoby matrix is calculated not at each step but only when required.

The program STIFSP [44] was developed for solution of similar problems. It is based on the program STIFF and on the method of equal roots suggested in [44]. The program was designed for operating with sparse matrices. $\Phi_{,y}$ and $\Phi_{,\dot{y}}$.

The program RADAU5 [53, 52] was designed for solving initial value problem for system of the form (4.8) with index bigger than two. In contrast to the above programs this program uses implicit methods of the Runge - Kutta type. Systems of nonlinear equations obtained are solved by the modified Newton's method. Comparison of the efficiencies of various programs was discussed in [53].

The program SINODE which uses certain numerical methods (such as Runge - Kutta and Adams methods) for solving the Cauchy problem for systems of equations of the form (4.8) with singular matrices was described in [11].

A numerical solution of the Cauchy problem for systems of differential - algebraic equations of the form

$$\dot{y}(t) = f(y, x, t),$$

$$x(t) = g(y, x, t)$$

was considered in [63]. It was suggested to combine implicit Euler method with simple iteration method or implicit Adams method with Newton's method.

Brief survey of approaches to numerical solutions of differential - algebraic equations shows certain achievements in this field. However, the difficulties mentioned above still remain.

The approach proposed in this chapter reduces some of these difficulties. As it will be seen later, the system of linear algebraic equations which appears at each step of integration process will be best - posed and, because of choice of the argument of the problem, the error will be less sensitive to the fast variations of the solution.

In the following section of this chapter the Cauchy problem for the differential - algebraic equations is considered from the standpoint of the parametric continuation method and the problem of choice of the best parameter is examined.

It is noted in the monograph [13] that program DASSL introduces a parameter into system of differential - algebraic equations. The program solves the system by means of the continuation method. However, the problem of choosing the best parameter of the solution continuation is not considered.

2.   The best argument for a system of differential - algebraic equations

We consider the Cauchy problem for a system of differential - algebraic equations

$$F(y, \dot{y}, x, t) = 0, \quad y(t_0) = y_0,$$

$$G(y, x, t) = 0, \qquad x(t_0) = x_0,$$

$$(4.9)$$

$$y(t) = (y_1(t), \ \ldots \ , y_n(t))^T, \qquad x(t) = (x_1(t), \ \ldots \ , x_m(t))^T,$$

$$F = (F_1, \ \ldots \ , F_n)^T, \qquad G = (G_1, \ \ldots \ , G_m)^T, \qquad t \in {}^1,$$

$$y_0 = (y_{10}, \ \ldots \ , y_{n0})^T, \qquad x_0 = (x_{10}, \ \ldots \ , x_{m0})^T,$$

$$\dot{y} = \frac{dy}{dt} = \left(\frac{dy_1}{dt}, \ \ldots \ , \ \frac{dy_n}{dt} \right)^T.$$

Here the vectors y_0, x_0 and value t_0 must be consistent, i.e., satisfy the system of equations $G(y_0, x_0, t_0) = 0$.

The integral of the problem (4.9)

$$f(y, x, t) = 0, \qquad f(y_0, x_0, t_0) = 0, \qquad\qquad (4.10)$$

$$f = (f_1, \ \ldots \ , \ f_{n+m})^T$$

specify a unique smooth integral curve K in the $(n+m+1)$ - dimensional Euclidean space ${}^{n+m+1}$. The process of its construction may by viewed as the process of continuation of the solution $y = y(t)$, $x = x(t)$ with respect to the parameter t. Such approach brings us the problem of choosing the best parameter of solution continuation of system (4.10) and, hence, the best argument of problem (4.9).

As in the chapters 1 and 2 we will introduce the best argument locally, i.e., in small vicinity of each point of the integral curve K. To find the best argument we introduce in the vicinity of the point considered such parameter μ that

$$d\mu = \alpha_i dy_i + \beta_j dx_j + \gamma dt, \quad i = \overline{1, \ n}, \quad j = \overline{1, \ m}. \qquad (4.11)$$

Here α_i, β_j, γ are components of the unit vector considered above $\alpha = (\alpha_1, \ \ldots \ , \ \alpha_n, \ \beta_1, \ \ldots \ , \ \beta_m, \ \gamma)^T \in R^{n+m+1}$ which specifies the direction in which the argument μ is measured. Here and later the summation in products with respect to repeating indexes in stipulated limits is assumed.

The functions $y_i(\mu)$, $x_j(\mu)$, $t(\mu)$ are assumed to be differentiable. Dividing equation (4.11) by $d\mu$ and differentiating the first of relations (4.10) with respect to μ we obtain the following continuation equations for the problem (4.9)

$$\alpha_i y_{i,\mu} + \beta_j x_{j,\mu} + \gamma t_{,\mu} = 1,$$

$$f_{,y_i} y_{i,\mu} + f_{,x_j} x_{j,\mu} + f_{,t} t_{,\mu} = 0. \qquad\qquad (4.12)$$

Here $y_{i,\mu} = dy_i/d\mu$, $f_{,y_i} = \partial f/\partial y_i$, ...

However, such approach is not constructive since integral (4.10) is unknown until problem (4.9) is solved.

The continuation equations can be obtained in another way. Let us linearize the vector function F with respect to $\dot{y}_i$ in a vicinity of certain value $\dot{y}_i = \dot{y}_i^*$ obtained, for example, at the previous step of the iterative process of integration procedure. Then

$$F^* + F_{,\dot{y}_i}^* (\dot{y}_i - \dot{y}_i^*) = 0, \qquad i = \overline{1, n}.$$

Here the vector functions F^* and $F_{,\dot{y}_i}^*$ are calculated at $\dot{y}_i = \dot{y}_i^*$.

Taking into account the first equation of system (4.12) and the equations $\dot{y}_i = y_{i,\mu}/t_{,\mu}$; $\dot{y}_i^* = y_{i,\mu}^*/t_{,\mu}^*$ and differentiating the vector function G with respect to μ we arrive at the continuation equations in the form

$$\alpha_i y_{i,\mu} + \beta_j x_{j,\mu} + \gamma t_{,\mu} = 1,$$

$$t_{,\mu}^* F_{,\dot{y}_i}^* y_{i,\mu} + \left(F^* t_{,\mu}^* - F_{,\dot{y}_i}^* y_{i,\mu}^* \right) t_{,\mu} = 0, \qquad (4.13)$$

$$G_{,y_i} y_{i,\mu} + G_{,x_j} x_{j,\mu} + G_{,t} t_{,\mu} = 0.$$

The integral curve for problem (4.9) can be constructed by integrating the system of ordinary differential equations obtained by resolving the continuation equations (4.13) with respect to derivatives with account for the initial conditions

$$y_i(0) = y_{i0}, \qquad x_j(0) = x_{j0}, \qquad t(0) = t_0. \qquad (4.14)$$

We assume here that the argument μ is measured from the initial point of the problem (4.9).

The conditionality of system (4.13) depends on the choice of the argument μ which, in turn, is determined by the vector α. The structure of this system completely coincides with the structure of the system (1.37) considered when proving the necessary and sufficient conditions for choosing the best parameter of solution continuation for a system of nonlinear algebraic or transcendental equations that contain a parameter. The parameter that ensures the best conditionality for the system of linear continuation equations is the arc length λ measured along the curve of solutions of the system (4.10) which in this case is an integral curve K for the problem (4.9) and the continuation parameter for system (4.10) is the argument of problem (4.9). We will name the argument $\mu = \lambda$, which ensures the best conditionality for the system of continuation equations (4.13), as the best argument. Note that the value of the determinant of the system divided by the product of the Euclidean norms of the matrix rows is taken as the measure of conditionality. It was shown in the proof of the theorem 1 that errors of numerical solution are minimal if the best argument is chosen.

In accordance with the Kramer's rule the solution of system (4.13) in this case can be represented in the form

$$\frac{dy_i}{d\lambda} = \frac{\Delta_i}{\Delta}, \qquad \frac{dx_j}{d\lambda} = \frac{\Delta_{n+j}}{\Delta}, \qquad \frac{dt}{d\lambda} = \frac{\Delta_{n+m+1}}{\Delta}, \qquad (4.15)$$

$$i = \overline{1, \, n}, \qquad j = \overline{1, \, m},$$

where Δ is the determinant of the system; $\Delta_k = (-1)^{k+1}\delta_k,$ ($k = \overline{1, \, n+m+1}$); δ_k is the determinant of the matrix that is obtained from the matrix of the last $n + m$ equations of the system by deleting its k - the column. These determinants satisfy the equation of the type (1.26) which can be written in the form

$$\Delta^2 = \Delta_k \Delta_k, \qquad (k = \overline{1, \, n+m+1}). \qquad (4.16)$$

This equation shows that the Euclidean norm of the right hand side of the system of ordinary differential equations (4.15) is always equal to one. If the argument λ is measured from the initial point of the problem (4.9) the initial conditions take the form (4.14).

Thus, on the basis of the results of chapter 1 we proved the following theorem.

Theorem 4. In order to formulate the Cauchy problem (4.9) for the system of differential - algebraic equations with respect to the best argument, it is necessary and sufficient to choose the arc length λ measured along the integral curve of the problem as this argument. In this case the problem (4.9) is transformed into the problem (4.15), (4.14) and the right hand sides of the problem (4.15) satisfy the relation (4.16).

In the following sections, which are based on this theorem, we will formulate an algorithm and describe programs of solutions for certain systems of differential - algebraic equations. We will test these programs on several benchmark problems. Previously, some approaches to the solution of such problems were considered in [106, 65, 107]. Here we describe a better algorithm.

3.   Explicit differential - algebraic equations

The Cauchy problem for the system of explicit differential - algebraic equations is given by

$$\frac{dy}{dt} = f(y, x, t), \quad y(t_0) = y_0,$$

$$G(y, x, t) = 0, \quad x(t_0) = x_0,$$

(4.17)

$$y : \ ^1 \longrightarrow \ ^n, \quad x : \ ^1 \longrightarrow \ ^m, \quad f : \ ^{n+m+1} \longrightarrow \ ^n,$$

$$G : \ ^{n+m+1} \longrightarrow \ ^m, \quad G(y_0, x_0, t_0) = 0.$$

It is obvious that the problem (4.17) is a special case of the problem (4.9). Let us formulate it with respect to the best argument λ assuming that functions $y = y(\lambda)$, $x = x(\lambda)$, $t = t(\lambda)$ are differentiable. Introduce the notations

$$\frac{dy}{d\lambda} = Y, \quad \frac{dx}{d\lambda} = X, \quad \frac{dt}{d\lambda} = T,$$

(4.18)

$$Y = (Y_1, \ \ldots \ , Y_n)^T, \quad X = (X_1, \ \ldots \ , X_m)^T.$$

Differentiating vector function G with respect to λ and taking into account relations (4.18) and meaning of the best argument, let us write system (4.17) in the form

$$Y_i \qquad\qquad - \quad f_i T \quad = \quad 0,$$

$$G_{,y_i} Y_i \ + \ G_{,x_j} X_j \ + \ G_{,t} T \quad = \quad 0,$$

(4.19)

$$Y_i Y_i \ + \quad X_j X_j \ + \quad TT \quad = \quad 1,$$

$$i = \overline{1, n}, \qquad j = \overline{1, m}.$$

Because of the last equation this system is nonlinear with respect to functions Y, X, T. However, it can be represented in linear form using the solution obtained at the previous $(k - 1)$ - the step. In order to achieve that we rewrite system (4.19) in the form

$$Y_i^{(k)} \qquad\qquad - \qquad f_i T^{(k)} \quad = \quad 0,$$

$$G_{,y_i} Y_i^{(k)} \ + \quad G_{,x_j} X_j^{(k)} \ + \quad G_{,t} T^{(k)} \quad = \quad 0,$$

(4.20)

$$Y_i^{(k-1)} Y_i^{(k)} \ + \ X_j^{(k-1)} X_j^{(k)} \ + \ T^{(k-1)} T^{(k)} \quad = \quad 1,$$

$$k = 1, 2, \quad \ldots$$

Hear $Z^{(k)} = (Y^{(k)}, X^{(k)}, T^{(k)})^T$ denotes a $n+m+1$ - dimensional vector. Because of the structure of system (4.20) this vector is tangential to the integral curve K of problem (4.17) at the point corresponding to the k - th step. Thus the last equation of system (4.20) is a scalar product of vectors $Z^{(k)}$ and $Z^{(k-1)}$ tangential to the integral curve at k - th and $(k-1)$ - th steps. This equation states that projection of the vector $Z^{(k)}$ on the direction of the unit vector $Z^{(k-1)}$ is equal to one (see Fig. 4.1). Replacing in (4.20) the unknown vector $Z^{(k)}$ with known vector $Z^{(k-1)}$ we ensure such local choice of an argument which is close to the best one.

Clearly, the vector $Z^{(k)}$ which satisfies the system of linear equations (4.20) is not, generally speaking, a unit vector, as required by the system (4.19). There-fore, after finding a solu-tion of system (4.20) the obtained vector $Z^{(k)}$ should be normalized by means of the formulas

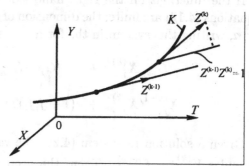

Figure 4.1.

$$Z_i^{*(k)} = \frac{Z_i^{(k)}}{\sqrt{Z_j^{(k)} Z_j^{(k)}}}, \qquad (4.21)$$

$$i, j = \overline{1, \, n+m+1}.$$

This yields a solution to system (4.19). Below we will omit the asterisk in (4.21).

Since the initial point is not usually a limiting point with respect to t, we can take initial approximation of the vector Z in the form

$$Z^{(0)} = (0, \, \ldots \, , \, 0, 1)^T. \qquad (4.22)$$

Assuming that the argument λ is counted from the initial point of the problem (4.17), the following algorithm for its solution can be proposed.

The solution of differential equations (4.18) satisfying to initial con-ditions

$$y(0) = y_0, \qquad x(0) = x_0, \qquad t(0) = t_0. \qquad (4.23)$$

is obtained.

Right hand sides of the system (4.18) are determined from the solution of system of linear equations (4.20) by Gauss elimination method. This solution is normalized by formulas (4.21).

Such approach allows not only to overcome the difficulties associated with vanishing of the Jacobian $G_{,x_j}$ but also to solve systems (4.17) in which the right hand sides f of differential equations become infinite at some points. To overcome difficulties associated with the latter case it is sufficient to rewrite, where it is possible, first n equations of system (4.19) in the form

$$Q_\alpha Y_\alpha - P_\alpha T = 0.$$

Here $\alpha = \overline{1, n}$ and summation by this index is not performed, functions Q_α, P_α are finite.

If the functions on the right hand side of the system of differential equations (4.17) are finite, the dimension of system (4.20) can be reduced by n, writing this system in the form

$$
\begin{aligned}
G_{,x_j} X_j^{(k)} \;+\; & (G_{,t} + G_{,y_i} f_i) T^{(k)} \;=\; 0, \\
X_j^{(k-1)} X_j^{(k)} \;+\; & (1 + f_i f_i) T^{(k-1)} T^{(k)} \;=\; 1.
\end{aligned}
\tag{4.24}
$$

Given a solution to system (4.24) the values $Y_i^{(k)}$ are defined by the formulas $Y_i^{(k)} = f_i T^{(k)}$. Then, the vector $Z^{(k)}$ should be normalized according to (4.21) and the values obtained should be used as the right hand sides for system (4.18).

These two approach were implemented in the programs DAE1EXG and DAE1EXP, written in FORTRAN. They solve the systems (4.20) and (4.24) correspondingly. The system (4.18) is integrated by the program PC1 and the system of linear equations is solved by Gauss elimination method.

In these programs the unknown functions y_i, x_j, t; $i = \overline{1, n}, j = \overline{1, m}$, are represented by a single array y_k, $k = \overline{1, n+m+1}$; $y_i = y_i$, $y_{n+j} = x_j$, $y_{n+m+1} = t$ and the coefficients of the system of linear equations (4.20) were given by the elements a_{kl} of a matrix of order $n+m+1$.

Examples.

As a first example we consider the problem [18]

$$
\begin{aligned}
\frac{dy}{dt} &= y + x^2, \\
\sqrt{y} - x &= 0,
\end{aligned}
\tag{4.25}
$$

$$y(0) = 4, \qquad x(0) = 2.$$

The problem has an analytical solution

$$y = 4e^{2t}, \qquad x = 2e^t.$$

The integration of problem (4.25) by program DA1EXG took 25% time more in comparison with program DA1EXP. The accuracy of computation was practically the same.

As a second example we consider the problem [13, 52]

$$\dot{y}_1 + \eta t \dot{y}_2 + (1 + \eta) y_2 = 0,$$

$$y_1 + \eta t y_2 = g(t), \qquad \dot{y} = \frac{dy}{dt}, \tag{4.26}$$

where $g(t)$ is a differentiable function, η is a number.

This problem has an exact solution

$$y_1 = g(t) + \eta t \dot{g}(t),$$

$$y_2 = -\dot{g}(t). \tag{4.27}$$

An error of computation was estimated by the formulas

$$\Delta_m = y_m^* - y_m, \qquad m = 1, 2, \tag{4.28}$$

where y_m^* is numerical solution and y_m is exact solution.

It is noted in [13, 52] that this problem is difficult for any numerical method. It's index is equal to 2 and numerical methods are often unstable for large indexes. For example, the solution of this problem by implicit Euler method are unstable for $\eta < -0.5$.

We apply the algorithm described above to the solution of the problem (4.26). After decreasing its index we use the method of j - continuation [11]. In order to do that, we differentiate the second equation (4.26) with respect to t twice. Taking into account the first equation we obtain a system

$$\dot{y}_1 + \eta t \dot{y}_2 + (1 + \eta) y_2 = 0,$$

$$\dot{y}_2 = -\ddot{g}(t). \tag{4.29}$$

We solve this problem with initial conditions

$$y_1(0) = g(0), \qquad y_2(0) = -\dot{g}(0). \tag{4.30}$$

The problem (4.29), (4.30) was solved for $g(t) = e^t$ and $\eta = 1, \eta = -2$ for $t \in [0, 1]$. With practically equal accuracy of computations the program DA1EXG took 30% more computation time. By means of the program DA1EXP the following results were obtained for $t = 1$. Here t_c is the computation time.

$$\eta = 1, \quad t_c = 57s, \quad \Delta_1 = 0.3 \cdot 10^{-2}, \quad \Delta_2 = -0.73 \cdot 10^{-3},$$

$$\eta = -2, \quad t_c = 68s, \quad \Delta_1 = -0.28 \cdot 10^{-2}, \quad \Delta_2 = -0.57 \cdot 10^{-3}.$$

Thus, it is better to use the program DA1EXP in a standard case.

Let us consider a differential equations which is nonlinear with respect to derivative

$$y - t - \dot{y} + \ln \dot{y} = 0, \qquad y(1) = e. \qquad (4.31)$$

Its analytical solution is $y = e^t$. The accuracy of computation is estimated as

$$\Delta = y - e^t, \qquad (4.32)$$

where y is a computed function.

The equation (4.31) also has the particular solution $y = t + 1$. That is why the equation is solved in the interval $1 \leq t \leq 2$ where the integral curve and the particular solution have no common points as well.

Let us transform the problem (4.31) to the form (4.17)

$$\frac{dy}{dt} = x, \qquad\qquad y(1) = e,$$

$$y - t - x + \ln x = 0, \quad x(1) = e.$$

This problem was solved by the program DA1EXP with initial step of integration of 0.001 and with accuracy of 10^{-5}. At the instant $t = 2$ the error estimated by (4.32) was equal to $\Delta = 0.28 \cdot 10^{-2}$. The initial step of integration was doubled three times.

4.   Implicit ordinary differential equations

Let us consider the problem

$$f(t, y_1, \ldots, y_n, \dot{y}_1, \ldots, \dot{y}_n) = 0,$$

$$y_i(t_0) = y_{i0}, \qquad\qquad i = \overline{1, n}, \qquad (4.33)$$

$$f = (f_1, \ldots, f_n)^T, \qquad\qquad \dot{y}_i = \frac{dy_i}{dt}.$$

If Jacoby matrix $\partial f / \partial \dot{y}$ is not singular then this system of implicit equations is a differential - algebraic system of index zero.

Introduction of new variables $x_i = \dot{y}_i$ transforms this problem into problem (4.17) in the extended space of variables. It appears that this transformation solves the problem (4.33), since the solution algorithm for problem (4.17) has been developed and described in the previous section.

However, the below example shows that such a transformation is not always justified and can even result in insurmountable computational difficulties.

We consider the Cauchy problem for the degenerate Van - der - Pol equation (4.4) written in the implicit form

$$(1 - y^2)\frac{dy}{dt} - y = 0, \qquad y(0) = 2. \tag{4.34}$$

The particular integral of this problem can be represented in the form

$$\ln\frac{y}{2} - \frac{y^2}{2} + 2 - t = 0. \tag{4.35}$$

The plot of this continuous function in the plane (y, t) is shown at Fig. 4.2 by the solid line.

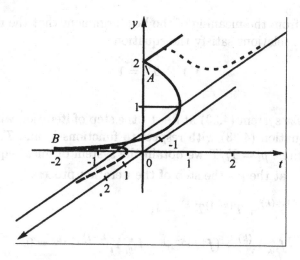

Figure 4.2.

Transforming problem (4.34) to the form (4.17) (this form was given in [52]) we obtain

$$\frac{dy}{dt} = x, \qquad\qquad y(0) = 2,$$
$$(1 - y^2)x - y = 0, \quad x(0) = -\frac{2}{3}. \tag{4.36}$$

The integral curve for this problem in the space of variables t, y, x is shown at Fig.4.2 by the dashed line. It has a discontinuity along the straight line $y = 1$, $t = 3/2 - \ln 2$. Therefore, an attempt to solve the problem (4.36) in the vicinity of this line by means of any numerical methods results in insurmountable computational difficulties.

This example shows that one should be careful when transforming (4.33) to problem (4.17) in the extended space of variables. Sometimes, however, such transition is justified although it increases the index of the system by one.

We consider algorithm for numerical solution of the problem (4.33) without its transformation to the form (4.17). Clearly, the problem (4.33) is a particular case of problem (4.9). Let us formulate problem (4.33) in terms of best argument.

Let y_i and t be functions of the best argument λ measured from the initial point of problem (4.33). Let us introduce the notations

$$\frac{dy_i}{d\lambda} = Y_i, \qquad \frac{dt}{d\lambda} = T, \qquad i = \overline{1, n}. \qquad (4.37)$$

It follows from the meaning of the best argument that the right hand sides of these relations satisfy the equation

$$Y_i Y_i + T^2 = 1. \qquad (4.38)$$

We linearize system (4.33) at $k-1$ - the step of iteration with respect to $\dot{y}_i$ and equation (4.38) with respect to functions Y_i and T. Then, in view of relations $y_i = Y_i/T$ we obtain the system of linear equations in $Y_i^{(k)}$ and $T^{(k)}$ at the k - the step of the iteration process

$$Y_i^{(k-1)} Y_i^{(k)} + T^{(k-1)} T^{(k)} = 1,$$

$$T^{(k-1)} f_{,\dot{y}_i}^* Y_i^{(k)} + (f^* T^{(k-1)} - f_{,\dot{y}_i}^* Y_i^{(k-1)}) T^{(k)} = 0. \qquad (4.39)$$

Here the asterisk marks the vector functions calculated for $\dot{y}_i = Y_i^{(k-1)}/T^{(k-1)}$. Whenever possible it is recommended to write system (4.39) in the form that does not include relations that tend to infinity and the terms containing Y_i and T in denominators.

If initial point is not a limiting point then the value of the vector $Z = (Y_1, \dots, Y_n, T)^T$ can be taken in the form

$$Z^{(0)} = (0, \dots, 0, 1). \qquad (4.40)$$

Thus, the problem reduces to the integration of the system of ordinary differential equations (4.37) that satisfy the initial conditions

$$y_i(0) = y_{i0}, \qquad t(0) = t_0. \qquad (4.41)$$

The right hand sides of system (4.37) are determined from the solution of system (4.39) normalized by formulas of the type (4.21).

Clearly, the last n equations of system (4.39) determine Newton - Raphson procedure, thus, this system of equations is solved until it converges to a given accuracy ε : $\|Z^{(k)} - Z^{(k-1)}\| < \varepsilon$. Note that if the system of ordinary differential equations (4.37) is solved by the program PC1, this condition is ensured by the predictor - corrector method. The solution calculated at a certain step of the integration process is taken as the initial approximation for the iteration procedure of the next step.

Obviously, system (4.39) takes the simplest form when the system of ordinary differential equations (4.33) is linear with respect to the derivatives $\dot{y}_i$, i.e., it is given by

$$a_{ij}(y_1, \ldots, y_n, t)\frac{dy_j}{dt} + g_i(y_1, \ldots, y_n, t) = 0, \qquad i, j = \overline{1, n}.$$

It was mentioned earlier that although the last equation in system (4.39) is approximate we obtain, after normalization, the solution of the nonlinear system

$$a_{ij}Y_j^{(k)} + a_{i\,n+1}T^{(k)} = 0,$$

$$Y_i^{(k)}Y_i^{(k)} + T^{(k)}T^{(k)} = 1, \qquad a_{i\,n+1} = g_i,$$

which does not depend on $Y_i^{(k-1)}$, $T^{(k-1)}$.

The discussed algorithm was used to design the program DE1ILN in which the problem (4.37), (4.41) is integrated by the program PC1 and the system (4.39) is solved by the Gauss elimination method.

Another algorithm can be proposed for solving the nonlinear problem (4.33). This algorithm does not involve the linearization of equations but requires an additional differentiation. Taking the equation $\dot{y}_i = Y_i/T$ into account, let us rewrite equation (4.33) in the form

$$F(t, y_1, \ldots, y_n, Y_1, \ldots, Y_n, T) = 0, \quad F = (F_1, \ldots, F_n)^T, \quad (4.42)$$

such that, whenever possible, the terms containing Y_i and T as the divisors and the relations tending to infinity are eliminated.

Differentiating equations (4.38) and (4.42) with respect to λ, we obtain the system of linear equations

$$F_{,Y_i} Y_i' + F_{,T} T' = -F_{,y_i} Y_i - F_{,t} T,$$

$$Y_i Y_i' + T T' = 0, \tag{4.43}$$

in the functions

$$\frac{dY_i}{d\lambda} = Y_i', \qquad \frac{dT}{d\lambda} = T', \qquad i = \overline{1,n}. \qquad (4.44)$$

Now, the problem is to integrate the system of ordinary differential equations (4.37), (4.44) that satisfy the initial conditions (4.41) and the following conditions

$$Y_i(0) = Y_{i0}, \qquad T(0) = T_0, \qquad i = \overline{1,n}. \qquad (4.45)$$

The right hand sides of equations (4.44) are determined from the solution of linear system (4.43) and initial values (4.45) of the functions Y_{i0} and T_0 are obtained from the following system of equations

$$F(t_0, y_{10}, \ \cdots \ , y_{n0}, \ Y_{10}, \ \cdots \ , Y_{n0}, \ T_0) = 0,$$

$$Y_{i0}Y_{i0} + T_0^2 = 0. \qquad (4.46)$$

This algorithm was implemented in the program DE1IN in which the system of differential equations is integrated by the program PC1 and the system of linear equations is solved by the Gauss elimination method.

Examples

As a first example we consider the numerical solution of problem (4.34). Computational error was estimated by substituting the computed solution into the left hand side of (4.35). The differential equation was integrated from point A to point B (Fig. 4.2), corresponding to $t = -2$.

The solution of this problem by DE1ILN took 28 dimensionless seconds. The error at $t = -2$ was equal to $\Delta = 0.55 \cdot 10^{-2}$. The initial value of the vector Z was taken in the form (4.40).

Under the same conditions the solution of this problem by the DE1IN program took 43 dimensionless seconds. The error was $\Delta = 0.27 \cdot 10^{-1}$. The initial values $Y_0 = -2/\sqrt{13}$, $T_0 = 3/\sqrt{13}$ were determined from the system of nonlinear equations

$$(1 - y_0^2)Y_0 - y_0 T_0 = 0,$$

$$Y_0^2 + T_0^2 = 1.$$

Note, that equation (4.34) can be written in the form resolved with respect to derivatives. After using λ – transformation the problem (4.34) takes the form

$$\frac{dy}{d\lambda} = -\frac{y}{\sqrt{(1-y^2)^2 + y^2}}, \quad y(0) = 2,$$

$$\frac{dt}{d\lambda} = -\frac{1-y^2}{\sqrt{(1-y^2)^2 + y^2}}, \quad t(0) = 0.$$

This problem was integrated in 14 dimensionless seconds. The error at $t = -2$ was equal to $\Delta = 0.55 \cdot 10^{-2}$.

As the second example we consider the solution of the problem (4.26) transformed to form (4.29). The solution of the problem for $t \in [0,1]$ by DE1TLN program led to the following results when $g(t) = e^t$

$$\eta = 1, \quad t_c = 39s, \quad \Delta_1 = 0.79 \cdot 10^{-3}, \quad \Delta_2 = 0.21 \cdot 10^{-3},$$

$$\eta = -2, \quad t_c = 42s, \quad \Delta_1 = 0.23 \cdot 10^{-2}, \quad \Delta_2 = -0.53 \cdot 10^{-3}.$$

Here computation errors were estimated by the formulas (4.28) for $t = 1$ and the initial value of vector Z was taken in the form (4.40). Integration of the problem by DE1IN led to the following results

$$\eta = 1, \quad t_c = 48s, \quad \Delta_1 = -0.93 \cdot 10^{-4}, \quad \Delta_2 = -0.33 \cdot 10^{-4},$$

$$\eta = -2, \quad t_c = 51s, \quad \Delta_1 = 0.79 \cdot 10^{-3}, \quad \Delta_2 = 0.21 \cdot 10^{-3}.$$

The initial values (4.45) were determined as a solution of nonlinear equation system at initial point

$$Y_1 + \eta t Y_2 + (1+\eta)y_2 T = 0,$$

$$Y_2 + \ddot{g}(t)T = 0,$$

$$Y_1^2 + Y_2^2 + T^2 = 1.$$

They were equal to

$$Y_{10} = \frac{(1+\eta)\dot{g}(0)}{Q}, \quad Y_{20} = -\frac{\ddot{g}(0)}{Q}, \quad T_0 = \frac{1}{Q},$$

$$Q = \left[1 + (1+\eta)^2 \dot{g}^2(0) + \ddot{g}^2(0)\right]^{\frac{1}{2}}.$$

Finally we consider the solution of the problem (4.31). After linearization with respect to $\dot{y}$, system (4.39) takes the form

$$\left(\frac{T^{(k-1)}}{Y^{(k-1)}} - 1\right) Y^{(k)} + \left[y - t + \ln\frac{Y^{(k-1)}}{T^{(k-1)}} - 1\right] T^{(k)} = 0,$$

$$Y^{(k-1)}Y^{(k)} + T^{(k-1)}T^{(k)} = 1. \tag{4.47}$$

This system involves the logarithmic function. Therefore, not every initial value of the vector Z is acceptable. For example, the vectors $(0,1)$; $(1,1)/\sqrt{2}$; $(1,\sqrt{3})/2$; $(1,2)/\sqrt{5}$ lead to uncertain results.

If we take the exact initial value of the vector Z in the form

$$Z^{(0)} = \frac{(e,\ 1)}{Q}, \qquad Q = (1+e^2)^{\frac{1}{2}}, \tag{4.48}$$

which is the solution of system

$$y - t - \frac{Y}{T} + \ln \frac{Y}{T} = 0,$$

$$Y^2 + T^2 = 1,$$

at initial point, then we obtain a reliable solution of the problem that does not depend on the initial value of Z if the solution of system (4.47) is accompanied by the normalization of the vector Z. Same is true if $Z^{(0)}$ is taken in the form $(\sqrt{3},1)/2$; $(2,1)/\sqrt{5}$, which is close to the exact initial value. If Z is not normalized reliable results are obtained only for exact value of $Z^{(0)}$ which is given by (4.48).

It took 23 dimensionless seconds to solve the problem (4.31) in the interval $t \in [1,2]$ by the DE1ILN program. The error (4.32) estimated at $t = 2$ was $\Delta = 0.29 \cdot 10^{-2}$.

Let us consider the solution of this problem by the DE1IN program. The equation (4.42) in this case takes the form

$$(y - t + \ln Y - \ln T)T - Y = 0.$$

The derivatives Y' and T' satisfy the system of linear equations

$$\left(\frac{T}{Y} - 1\right)Y' + (y - t + \ln Y - \ln T - 1)T' = (T - Y)T,$$

$$YY' + TT' = 0.$$

When the system of differential equations (4.44) is integrated the initial conditions should be taken in the form

$$Y(0) = \frac{e}{Q}, \qquad T(0) = \frac{1}{Q}$$

in accordance with (4.46).

The computation time was 27 dimensionless seconds. The error at the instant $t = 2$ was equal to $\Delta = 0.15 \cdot 10^{-3}$.

The following conclusions result from our numerical experiments.

1. The initial value of the vector Z for the DE1ILN program cannot be always taken in the form (4.40). Sometimes its components should

be varied. They also might be chosen close to the exact values obtained by solving the system (4.46).

2. Normalization of solution of the linear equations system (4.39) allows one to avoid difficulties associated with the use of approximate value of the vector $Z^{(0)}$.

3. Transformation of the system of implicit differential equations (4.33) to the system of explicit differential - algebraic equations (4.17) results in increase of the computation time and reduces the accuracy of computations.

In comparison with the DE1ILN program the DE1IN program requires more computational time but, as a rule, provides higher accuracy. Moreover, the DE1IN program has the following serious drawbacks:

— the transition to system (4.43) involve additional differentiation which can be a rather difficult problem;

— the order of integrated system of differential equations (4.37), (4.44) is twice the order of the original system which may result in a large computation time for a high order system of implicit equations;

— the program requires exact initial conditions (4.45) which is a solution of nonlinear system (4.46).

As another example, we consider numerical integration of implicit system of Euler's kinematic equations (2.40). The matrix of the system tends to be singular in vicinity of the value $\theta = 0$ and solving this system by DE1ILN program with conditions $\omega_1 = -100$, $\omega_2 = 1$, $\omega_3 = 0$, $\psi = \varphi = t = 0$, $\theta = \pi/100$ fails.

After λ – transformation of the system (2.40) it can be written in the form

$$
\begin{pmatrix}
\sin\theta \sin\varphi & 0 & \cos\varphi & -\omega_1 \\
\sin\theta \cos\varphi & 0 & -\sin\varphi & -\omega_2 \\
\cos\theta & 1 & 0 & -\omega_3 \\
z_1 & z_2 & z_3 & z_4
\end{pmatrix}
\begin{pmatrix}
\psi_{,\lambda} \\
\varphi_{,\lambda} \\
\theta_{,\lambda} \\
t_{,\lambda}
\end{pmatrix}
=
\begin{pmatrix}
0 \\
0 \\
0 \\
1
\end{pmatrix}.
$$

This system of equations is solved successfully by the DE1ILN program both for conditions $\omega_1 = -100$, $\omega_2 = 1$, $\omega_3 = 0$, $\psi = \varphi = t = 0$, $\theta = \pi/100$, as well as for the latter condition $\theta = 0$, if initial value of vector Z is taken in the form $Z^{(0)} = (1, 0, 0, 0)$.

Note that in the same way the following system of kinematic equations for airplane angles can be solved too [73]

$$
\begin{pmatrix}
\sin\vartheta & 0 & 1 \\
\cos\vartheta \cos\gamma & \sin\gamma & 0 \\
-\cos\vartheta \sin\gamma & \cos\gamma & 0
\end{pmatrix}
\begin{pmatrix}
\psi_{,t} \\
\vartheta_{,t} \\
\gamma_{,t}
\end{pmatrix}
=
\begin{pmatrix}
\omega_1 \\
\omega_2 \\
\omega_3
\end{pmatrix}.
$$

Here ψ is angle of yaw, ϑ is pitch angle and γ is angle of bank.

The matrix of this system becomes singular at $\vartheta = \pi/2$. After λ – transformation the latter system takes the form

$$\begin{pmatrix} \sin\vartheta & 0 & 1 & -\omega_1 \\ \cos\vartheta\cos\gamma & \sin\gamma & 0 & -\omega_2 \\ -\cos\vartheta\sin\gamma & \cos\gamma & 0 & -\omega_3 \\ z_1 & z_2 & z_3 & z_4 \end{pmatrix} \begin{pmatrix} \psi_{,t} \\ \vartheta_{,t} \\ \gamma_{,t} \\ t_{,\lambda} \end{pmatrix} = \begin{pmatrix} 0 \\ 0 \\ 0 \\ 1 \end{pmatrix}.$$

5.   Implicit differential - algebraic equations

Implicit differential - algebraic equations define the Cauchy problem in the form

$$f(t, y_1, \ldots, y_n, x_1, \ldots, x_m, \dot{y}_1, \ldots, \dot{y}_n) = 0,$$

$$G(t, y_1, \ldots, y_n, x_1, \ldots, x_m) = 0,$$

(4.49)

$$y_i(t_0) = y_{i0}, \quad x_j(t_0) = x_{j0}, \quad i = \overline{1, n}, \quad j = \overline{1, m},$$

$$f = (f_1, \ldots, f_n)^T, \quad G = (G_1, \ldots, G_m)^T,$$

$$G(t_0, y_{10}, \ldots, y_{n0}, x_{10}, \ldots, x_{m0}) = 0, \quad \dot{y}_i = \frac{dy_i}{dt}.$$

By introducing new variables $z_i = \dot{y}_i$ one can transform this problem into the explicit problem (4.17) in the extended space of variables. It was shown above, however, that such an approach may result in computational difficulties.

We consider an algorithm for solving problem (4.49) without transformation of it to the form (4.17). For problem (4.49) which coincide with the problem (4.9) we proved Theorem 4 which determines the best argument λ. Let us formulate the problem with respect to this argument.

Let $y_i = y_i(\lambda)$, $x_j = x_j(\lambda)$, $t = t(\lambda)$ be differentiable functions of the argument λ that is counted along the integral curve from the initial point of the Cauchy problem (4.49).

Taking into account the notation (4.18) and the meaning of the best argument we obtain the equations

$$Y_i Y_i + X_j X_j + T^2 = 1. \tag{4.50}$$

Let us linearize the system (4.49) and the equation (4.50) with respect to derivatives $\dot{y}_i$ and quadratic terms correspondingly. We also differentiate the vector function G with respect to λ. Thus, we obtain the system of equations that are linear with respect to the functions

Y_i, X_j, T calculated at the k-th step of the iteration process

$$T^{(k-1)} f_{,\dot{y}_i}^* Y_i^{(k)} + \left(f^* T^{(k-1)} - f_{,\dot{y}_i}^* Y_i^{(k-1)} \right) T^{(k)} = 0,$$

$$G_{,y_i} Y_i^{(k)} + G_{,x_j} X_j^{(k)} + G_{,t} T^{(k)} = 0, \qquad (4.51)$$

$$Y_i^{(k-1)} Y_i^{(k)} + X_j^{(k-1)} X_j^{(k)} + T^{(k-1)} T^{(k)} = 1.$$

Here the asterisk marks the functions calculated at the previous step, i.e., for

$$\dot{y}_i = Y_i^{(k-1)} / T^{(k-1)}.$$

If initial point is not a limiting point, initial value of the vector $Z = (Y_1, \ldots, Y_n, X_1, \ldots, X_m, T)^T$ can be taken in the form (4.22). Now the problem is to integrate a system of ordinary differential equations (4.18) the right hand sides of which are obtained by solving the system of linear equations (4.51) with the help of Newton – Raphson method followed by normalization of the solution using formulas (4.21). Since argument λ is measured from the initial point of problem (4.49) the initial conditions for system (4.18) have the form

$$y_i(0) = y_{i0}, \quad x_j(0) = x_{j0}, \quad t(0) = t_0, \quad i = \overline{1, n}, \quad j = \overline{1, m}. \qquad (4.52)$$

Clearly, if system (4.49) is linear with respect to the derivatives $\dot{y}_i$ then the solution of (4.51), which is obtained by means of this approach, does not require iterative improvement, does not depend on the solution found at the previous step and satisfies the relation (4.50).

This algorithm was implemented in the DA1ILN program in which a system of differential equations is integrated by PC1 program and system of linear equations is solved by Gauss elimination method.

Note, when equations of problems (4.33), (4.49) are nonlinear with respect to the derivatives $\dot{y}_i$ then the procedure of prediction - correction method in PC1 program provides more precise iterative definition of the solution of the linearized systems (4.39), (4.51).

Another algorithm for solving problem (4.49) is as follows. Taking into account the relations $\dot{y}_i = Y_i/T$ and following the rules formulated in the previous section we rewrite the first vector equation of the problem in the form

$$F(t, y_1, \ldots, y_n, x_1, \ldots, x_m, Y_1, \ldots, Y_n, T) = 0. \qquad (4.53)$$

Let us differentiate equations (4.50) and (4.53) with respect to λ once and with respect to vector function G of the system (4.49) twice. Then we obtain the following system of linear equations for the derivatives of Y_i, X_j and T

$$F_{,Y_i} Y_i' \qquad\qquad + \quad F_{,T} T' \quad = \quad -(F_{,y_i} Y_i + F_{,x_j} X_j + F_{,t} T),$$

$$G_{,y_i} Y_i' + G_{,x_j} X_j' + G_{,t} T' \quad = \quad -(G'_{,y_i} Y_i + G'_{,x_j} X_j + G'_{,t} T),$$

$$Y_i Y_i' \quad + \quad X_j X_j' \quad + \quad TT' \quad = \quad 0.$$

$$(4.54)$$

Here prime denotes differentiation with respect to λ

$$\frac{dY_i}{d\lambda} = Y_i', \qquad \frac{dX_j}{d\lambda} = X_j', \qquad \frac{dT}{d\lambda} = T'. \qquad (4.55)$$

Thus, the problem is to solve the system of ordinary differential equations (4.18), (4.55) satisfying initial conditions (4.52) and the following conditions

$$Y_i(0) = Y_{i0}, \qquad X_j(0) = X_{j0}, \qquad T(0) = T_0,$$

$$i = \overline{1, n}, \qquad j = \overline{1, m}. \qquad (4.56)$$

The right hand sides of equations (4.55) satisfy the system of linear equations (4.54) and initial conditions (4.56) are determined as a solutions of nonlinear equations

$$F(t_0, y_{10}, \ldots, y_{n0}, x_{10}, \ldots, x_{m0}, Y_{10}, \ldots, Y_{n0}, T_0) = 0,$$

$$G^0_{,y_i} Y_{i0} + G^0_{,x_j} X_{j0} + G^0_{,t} T_0 = 0,$$

$$Y_{i0} Y_{i0} + X_{j0} X_{j0} + T_0^2 = 1,$$

$$(4.57)$$

where the superscript zero in the vector function G means that the derivative is calculated at the initial point of problem (4.49).

This algorithm was implemented in the DA1IN program in which a system of ordinary differential equations is integrated by the PC1 program and the system of linear equations is solved by the Gauss elimination method.

Examples.

We consider the Cauchy problem for the system

$$(1 - y^2)\frac{dy}{dt} - y = 0,$$

$$x^2 + 4(y - 0.5)^2 + t^2 = 3^2, \qquad (4.58)$$

$$y(0) = 2, \qquad x(0) = 0.$$

The integral curve of the problem is a line of intersection of the cylinder (4.35) with the ellipsoid given by second equation of system (4.58).

Error of computations can be estimated by substituting the computed solution into the right hand sides of formulas

$$\Delta_1 = \ln \frac{y}{2} - \frac{y^2}{2} - t + 2,$$

$$\Delta_2 = 4(y - 0.5)^2 + x^2 + t^2 - 9. \qquad (4.59)$$

The problem was solved from zero to $t = -2$ by means of the DA1ILN program with initial step of integration along λ equal to 0.001 and with an accuracy 10^{-5}. The execution time t_c was 138 dimensionless seconds. Errors (4.59) were equal to $\Delta_1 = 0.22 \cdot 10^{-2}$, $\Delta_2 = -0.96 \cdot 10^{-2}$.

The system of linear equations (4.51) in this case takes the form

$$(1 - y^2)Y^{(k)} \qquad\qquad - \qquad yT^{(k)} = 0,$$

$$4(y - 0.5)Y^{(k)} + \qquad xX^{(k)} + \qquad tT^{(k)} = 0, \qquad (4.60)$$

$$Y^{(k-1)}Y^{(k)} + X^{(k-1)}X^{(k)} + T^{(k-1)}T^{(k)} = 1.$$

If the initial value of the vector Z for this system is taken in the form (4.22) then we obtain a degenerate system of equations. However, the system was solved equally successfully both for $Z^{(0)} = (0, 1, 0)$ and $Z^{(0)} = (1, 1, 1)/\sqrt{3}$.

Note that for the same accuracy of computations the execution time increases to 163 dimensionless seconds if derivatives (4.15) are calculated by Cramer's rule taking into account relations (4.16).

The solution of problem (4.58) by means of the DA1IN program took $t_c = 163$ s. with errors $\Delta_1 = -0.3 \cdot 10^{-1}$ and $\Delta_2 = 0.13 \cdot 10^{-1}$.

In this case the system of linear equations (4.54) takes the form

$$(1 - y^2)Y' \qquad\qquad - \quad yT' \;=\; 2yY^2 + YT,$$

$$4(y - 0.5)Y' \;+\; xX' \;+\; tT' \;=\; -4Y^2 - X^2 - T^2,$$

$$YY' \;+\; XX' \;+\; TT' \;=\; 0.$$

The initial values for the system of differential equations (4.55) were determined by solving equations (4.60) at the initial point of problem (4.58)

$$Y^{(k-1)} = Y^{(k)} = Y = Y_0,$$

$$X^{(k-1)} = X^{(k)} = X = X_0, \qquad\qquad (4.61)$$

$$T^{(k-1)} = T^{(k)} = T = T_0.$$

They were equal to

$$Y_0 = Y(0) = 0, \qquad X_0 = X(0) = 1, \qquad T_0 = T(0) = 0.$$

To make first equation (4.58) dependent on the second one we substitute the expression $1 - y^2$ obtained from the latter into the former. Then this equation takes the form

$$(x^2 + t^2 - 4y - 4)\frac{dy}{dt} - 4y = 0.$$

The solution of problem (4.58) with this first equation led to the following results:

$t_c = 141$ s. $\Delta_1 = 0.45\cdot 10^{-2}$, $\Delta_2 = -0.95\cdot 10^{-2}$ for DA1ILN program; and $t_c = 167$ s. $\Delta_1 = 0.49\cdot 10^{-1}$, $\Delta_2 = -0.13\cdot 10^{-1}$ for DA1IN program.

Note that for any numerical method of solving of ordinary differential equations the problem under consideration is difficult. In addition to the above mentioned limiting point that lies on the straight line $y = 1$, $t = 3/2 - \ln 2$ there is a singularity at the initial point. At this point the derivative $G_{,x} = 2x$ is equal to zero that makes the determination of function X impossible by usual approach.

As another example let us consider the problem which is nonlinear with respect to the derivative.

$$y - t - \dot{y} + \ln\dot{y} = 0,$$

$$y - x^2 - t^2 = 0, \qquad\qquad (4.62)$$

$$y(1) = e, \qquad x(1) = \sqrt{e} - 1.$$

The integral curve of this problem is obtained by the intersection of the cylinder $y - e^t = 0$ with the paraboloid given by the second equation of system (4.62). The error of computations is estimated by substitution of the computed solution into the right hand sides of the formulas

$$\Delta_1 = y - e^t, \qquad \Delta_2 = y - x^2 - t^2. \tag{4.63}$$

It took 38 s. to solve the problem (4.62) by the DA1ILN program within the interval $t \in [1, 2]$ with the initial step of integration along λ equal to 0.001 and with the accuracy 10^{-5}. The errors (4.63) at $t = 2$ were equal to $\Delta_1 = 0.29 \cdot 10^{-2}$ and $\Delta_2 = 0.9 \cdot 10^{-3}$.

The system of linearized equations (4.51) has the form

$$\left(\frac{T^{(k-1)}}{Y^{(k-1)}} - 1 \right) Y^{(k)} + \left(y - t + \ln \frac{Y^{(k-1)}}{T^{(k-1)}} - 1 \right) T^{(k)} = 0,$$

$$Y^{(k)} - 2xX^{(k)} - 2tT^{(k)} = 0,$$

$$Y^{(k-1)}Y^{(k)} + X^{(k-1)}X^{(k)} + T^{(k-1)}T^{(k)} = 1.$$

We used both exact value of vector $Z^{(0)} = (Y_0, X_0, T_0)^T$ where values

$$Y_0 = \frac{2\sqrt{e-1}}{Q}, \qquad X_0 = \frac{e-2}{eQ}, \qquad T_0 = \frac{Y_0}{e}, \tag{4.64}$$

$$Q = \sqrt{4e - 3},$$

were determined as a solution of the system

$$e - 1 - \frac{Y_0}{T_0} + \ln \frac{Y_0}{T_0} = 0,$$

$$Y_2 - 2\sqrt{e-1}X_0 - 2T_0 = 0,$$

$$Y_0^2 + X_0^2 + T_0^2 = 1,$$

and approximate value $Z^{(0)} = (2, 2, 1)/3$. The results were the same in both cases.

The solution of problem (4.62) by means the DA1IN program took 43 s. The errors at $t = 2$ were equal to $\Delta_1 = 0.15 \cdot 10^{-3}$ and $\Delta_2 = 0.68 \cdot 10^{-4}$.

The values of derivatives Y', X', T' were found by solving the system of linear algebraic equations

$$\left(\frac{T}{Y} - 1 \right) Y' + \left(y - t + \ln \frac{Y}{T} - 1 \right) T' = T^2 - YT,$$

$$Y' - 2xX' - 2tT' = 2(X^2 + T^2),$$

$$YY' + XX' + TT' = 0.$$

The system of differential equations (4.55) was solved with initial conditions (4.64).

To make the first equation of system (4.62) dependent on the second one we took the former in the form

$$x^2 + t^2 - t - \dot{y} + \ln \dot{y} = 0.$$

In the case the following results were obtained: For the DA1ILN program executive time was equal to 39 s., The errors were $\Delta_1 = 0.21 \cdot 10^{-2}$, $\Delta_2 = 0.91 \cdot 10^{-3}$. For the DA1IN executive time was equal to 44 s., the errors were $\Delta_1 = 0.79 \cdot 10^{-3}$, $\Delta_2 = 0.51 \cdot 10^{-4}$.

This problem also was transformed to the form

$$\frac{dy}{dt} = x_2,$$

$$x_1^2 + t^2 - t - x_2 + \ln x_2 = 0,$$

$$y - x_1^2 - t^2 = 0,$$

$$y(1) = e, \qquad x_1(1) = \sqrt{e - 1}, \qquad x_2(1) = e,$$

and solved by means of the DA1EXP program. The system of equations (4.20) had the form

$$Y^{(k)} - x_2 T^{(k)} = 0,$$

$$2x_1 X_1^{(k)} + \left(\frac{1}{x_2} - 1\right) X_2^{(k)} + (2t - 1)T^{(k)} = 0,$$

$$Y^{(k)} - 2x_1 X_1^{(k)} - 2t T^{(k)} = 0,$$

$$Y^{(k-1)}Y^{(k)} + X_1^{(k-1)}X_1^{(k)} + X_2^{(k-1)}X_2^{(k)} + T^{(k-1)}T^{(k)} = 1.$$

The initial vector Z was taken in the form (4.22). The execution time was 43 s. and the computational errors were equal to $\Delta_1 = 0.28 \cdot 10^{-2}$ and $\Delta_2 = 0.99 \cdot 10^{-3}$.

As one more example we consider solving of the Cauchy problem for differential – algebraic equation with closed integral curve

$$(2P - b^2)y\frac{dy}{dt} = b^2 t - P(2t - a),$$

$$4y^2 + x^2 + t^2 - 16 = 0, \qquad\qquad (4.65)$$

$$y(0) = b, \qquad x(0) = 2\sqrt{4 - b^2},$$

where $P = P(y, t) = y^2 + t^2 - at$; a, b are given numbers.

The integral curve of the problem is a result of intersection of the cylinder the directrix of which is determined by equation $P^2(y,t) - b^2(y^2 + t^2) = 0$ with ellipsoid given by the second equation of system (4.65).

In order for the problem to be connected and taking into account the equation of ellipsoid the expression of function $P(y,t)$ was written in the form $P = P(y,t,x) = 16 - x^2 - 3y^2 - at$.

The following relation were used for estimating errors

$$\Delta_1 = (y^2 + t^2 - at)^2 - b^2(y^2 + t^2),$$
$$\Delta_2 = 4y^2 + x^2 + t^2 - 16,$$

(4.66)

The computed values of functions were substituted in those equations.

The problem was solved by DA1ILN program with accuracy 10^{-5} and the initial integral step along λ equal to 10^{-2} for $a = 1$ and $b = 1.5$.

The whole integral curve was computed in 160 s. At the end of the loop the errors computed by formulas (4.66) were equal to $\Delta_1 = 0.036$, $\Delta_2 = -0.042$.

The system of equations (4.51) has the form

$$(2P - b^2)yY^{(k)} - [b^2t - P(2t - a)]T^{(k)} = 0,$$

$$4yY^{(k)} + xX^{(k)} + tT^{(k)} = 0,$$

(4.67)

$$Y^{(k-1)}Y^{(k)} + X^{(k-1)}X^{(k)} + T^{(k-1)}T^{(k)} = 1.$$

The initial value of vector Z was taken as $Z^{(0)} = (1,1,1)/\sqrt{3}$.

The problem was solved by the DA1IN program. The execution time was equal to 255 s. and the errors were equal to $\Delta_1 = -0.012$ and $\Delta_2 = -0.037$.

The system of equations (4.54) had the form

$$(2P - b^2)yY' - [b^2t - P(2t - a)]T' =$$

$$= -[2P'y + (2P - b^2)Y]Y + [b^2T - P'(2t - a) - 2PT]T,$$

$$4yY' + xX' + tT' = -4Y^2 - X^2 - T^2,$$

$$YY' + XX' + TT' = 0,$$

where $P' = -2xX - 6yY - aT$.

The system of differential equations (4.18), (4.55) was integrated with initial conditions

$$y(0) = b, \quad x(0) = 2\sqrt{4 - b^2}, \quad t(0) = 0,$$

$$Y(0) = \frac{a\sqrt{4 - b^2}}{Q}, \quad X(0) = -\frac{2ab}{Q}, \quad T(0) = Y(0)\frac{b}{a},$$

$$Q = \sqrt{4(a^2 + b^2) + 3a^2b^2 - b^4},$$

in which the values of the functions Y, X, T were found as solutions of system of equations (4.67) with conditions (4.61).

The results of this section support the conclusions drawn in the previous section. As we suggested, increase of the problem dimensionality leads to substantial increase of execution time for DA1IN, although the accuracy of computation for DA1IN is about the same as that for DA1ILN.

Special attention must be given to the use of the DA1EXP program for solving problems that are nonlinear with respect to derivatives $\dot{y}_i$. Although the execution time for this program is a little greater than for DA1ILN program it provides practically the same accuracy. However, DA1EXP program does not require linearization of the equations. Hence, if transformation of the problem to the problem in extended space of variables does not lead to confusion, it seems reasonable to use the DA1EXP program.

One more point should be noted. It was shown in [13] that because of the ill conditionality the error is concentrated in the algebraic variable rather than in the differentiable one. It is proposed to use this fact for checking the accuracy of computations. The analysis of the numerical results obtained shows that both above errors are close to each other. This may be associated with the fact that the computational process is well conditioned.

As a more complex problem we consider the model problem of breaking out the canopy of axially symmetric parachute under given pressure p. The aerodynamic influence of strings is neglected. The tension of strings and frame bands obeys Hooke's law

$$N = \begin{cases} E\dfrac{dS - ds}{ds} = E(l - 1), & l = \dfrac{dS}{ds} > 1, \\ 0, & l \leq 1, \end{cases}$$

where N is a force; E is the modulus of elasticity, which has the dimension of a force; dS and ds are elements of the arc in deformed and undeformed state respectively.

Let us denote the mass of the string (or the radial frame band with the the fabric adjacent to it together with frame elements) per unit length in undeformed and deformed state as m_0 and m respectively. Then taking into account dissipative force, the equations of motion of the axially symmetric parachute takes the form [97, 9]

$$m_0\frac{\partial^2 x}{\partial t^2} = \frac{\partial}{\partial s}\left(\frac{N}{l}\frac{\partial x}{\partial s}\right) - \varepsilon l\frac{\partial x}{\partial t} - p\frac{2\pi r}{M}\frac{\partial r}{\partial s}\delta,$$

$$m_0\frac{\partial^2 r}{\partial t^2} = \frac{\partial}{\partial s}\left(\frac{N}{l}\frac{\partial r}{\partial s}\right) - \varepsilon l\frac{\partial r}{\partial t} + p\frac{2\pi r}{M}\frac{\partial x}{\partial s}\delta, \qquad (4.68)$$

$$\left(\frac{\partial x}{\partial s}\right)^2 + \left(\frac{\partial r}{\partial s}\right)^2 = l^2.$$

Here x, r are axis of fixed system of coordinate (Fig. 4.3) with its origin at the top of the parachute; t is time; ε is the dissipation coefficient; M is the number of strings; parameter $\delta = 1$ for band of radial frame and $\delta = 0$ for strings.

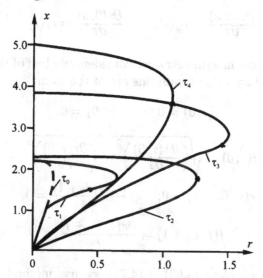

Figure 4.3.

We introduce the dimensionless quantities

$$s^* = \frac{s}{R_0}, \qquad L_c^* = \frac{L}{R_0}, \qquad x^* = \frac{x}{R_0}, \qquad r^* = \frac{r}{R_0},$$

$$\tau = \frac{t}{R_0}\sqrt{\frac{E}{m}}, \qquad e^* = \frac{\varepsilon R_0}{\sqrt{Em}}, \qquad p^* = \frac{2\pi R_0^2}{ME}p,$$

where m_c is the mass of the string per unit length; L_c is the length of string; R_0 is the radius of cutted out canopy of the parachute.

Taking into account the introduced notations, we write equations (4.68) in the form

$$\frac{\partial^2 x}{\partial \tau^2} = \frac{m}{m}\left[\nu\left(\frac{1}{l^2}\frac{\partial l}{\partial s}\frac{\partial x}{\partial s} + \left(1 - \frac{1}{l}\right)\frac{\partial^2 x}{\partial s^2}\right) - \varepsilon l\frac{\partial x}{\partial \tau} - p\frac{\partial r}{\partial s}r\delta\right],$$

$$\frac{\partial^2 r}{\partial \tau^2} = \frac{m}{m}\left[\nu\left(\frac{1}{l^2}\frac{\partial l}{\partial s}\frac{\partial r}{\partial s} + \left(1 - \frac{1}{l}\right)\frac{\partial^2 r}{\partial s^2}\right) - \varepsilon l\frac{\partial r}{\partial \tau} + p\frac{\partial x}{\partial s}r\delta\right], \qquad (4.69)$$

$$\left(\frac{\partial x}{\partial s}\right)^2 + \left(\frac{\partial r}{\partial s}\right)^2 = l^2.$$

In these equations the asterisk by the dimensionless values is omitted; $\nu = E_c/E$ if $l > 1$ and $\nu = 0$ if $l \leq 1$; E_c is the modulus of elasticity of string.

Initial conditions for system (4.69) are taken in the form

$$x(0, s) = x_0(s), \qquad r(0, s) = r_0(s),$$

$$\frac{\partial x(0, s)}{\partial \tau} = v_x(s), \qquad \frac{\partial r(0, s)}{\partial \tau} = v_r(s). \qquad (4.70)$$

Taking into account symmetry, we consider only half of the parachute. Thus, the boundary value conditions are of the form

$$x(\tau, 0) = 0, \qquad r(\tau, 0) = 0,$$

$$l(\tau, 0) = \sqrt{\left(\frac{\partial x(\tau, 0)}{\partial s}\right)^2 + \left(\frac{\partial r(\tau, 0)}{\partial s}\right)^2},$$

$$r(\tau, L_c + 1) = 0, \qquad \frac{\partial x(\tau, L_c + 1)}{\partial s} = 0, \qquad (4.71)$$

$$l(\tau, L_c + 1) = \frac{\partial r(\tau, L_c + 1)}{\partial s}.$$

To solve the problem (4.69) – (4.71) we use method of lines with respect to s coordinate. According to this method the dimensionless initial length of the parachute $L_c + 1$ is divided into $(n - 1)$ equal parts with the step $h = (L_c + 1)/(n - 1)$. The derivatives with respect to s are approximated to the second order of accuracy by central differences. For the function $Z(s)$ at the k - th point $(1 < k < n)$ the derivatives take the form

$$\frac{\partial Z}{\partial s} = \frac{Z_{k+1} - Z_{k-1}}{2h}, \qquad \frac{\partial^2 Z}{\partial s^2} = \frac{Z_{k+1} - 2Z_k + Z_{k+1}}{h^2}. \qquad (4.72)$$

The values of derivatives in the boundary conditions are determined as follows. Let us write Taylor series (limiting by terms of second order accuracy with respect to h) for the function $Z(s)$ in the vicinity of point $(Z = Z_1)$.

$$Z_2 = Z_1 + h\frac{\partial Z}{\partial s} + \frac{h^2}{2}\frac{\partial^2 Z}{\partial s^2},$$

$$Z_3 = Z_1 + 2h\frac{\partial Z}{\partial s} + \frac{4h^2}{2}\frac{\partial^2 Z}{\partial s^2}.$$

Here $Z(s)$ is a function of $x_i(s)$ or $r_i(s)$.

Solving this system with respect to derivatives we obtain that at the origin of coordinate system

$$\frac{\partial Z}{\partial s} = \frac{4Z_2 - 3Z_1 - Z_3}{2h}.$$

Taking into account that $\partial x/\partial s = 0$ we obtain expression $x_n = (4x_{n-1} - x_{n-2})/3$ in the center of canopy.

Using finite difference approximations we transform the problem (4.69) – (4.71) to the system of differential – algebraic equations which can be represented in the form

$$\frac{dx_i}{d\tau} = X_i, \qquad \frac{dX_i}{d\tau} = f_{x_i}, \qquad \frac{dr_i}{d\tau} = R_i, \qquad \frac{dR_i}{d\tau} = f_{r_i},$$

$$\left(\frac{x_{i+1} - x_{i-1}}{2h}\right)^2 + \left(\frac{r_{i+1} - r_{i-1}}{2h}\right)^2 = l_i^2, \qquad i = \overline{2, \, n-1}. \tag{4.73}$$

Here f_{x_i}, f_{r_i} are right hand sides of the first two equations of system (4.69) written at i - th point with the help of formulas (4.72).

Now we formulate the initial – value problem for system (4.73) with respect to the best argument λ. After differentiation of algebraic relations with respect to λ, the system (4.73) can be transformed to the form

$$\frac{dx_i}{d\lambda} = X_i\frac{d\tau}{d\lambda}, \qquad \frac{dX_i}{d\lambda} = f_{x_i}\frac{d\tau}{d\lambda},$$

$$\frac{dr_i}{d\lambda} = R_i\frac{d\tau}{d\lambda}, \qquad \frac{dR_i}{d\lambda} = f_{r_i}\frac{d\tau}{d\lambda},$$

$$\frac{dl_i}{d\lambda} = \frac{1}{l_i}\left(\frac{x_{i+1} - x_{i-1}}{2h} \cdot \frac{X_{i+1} - X_{i-1}}{2h} + \right.$$

$$\left. +\frac{r_{i+1} - r_{i-1}}{2h} \cdot \frac{R_{i+1} - R_{i-1}}{2h}\right)\frac{d\tau}{d\lambda} = f_{l_i}\frac{d\tau}{d\lambda}.$$

Taking into account that the best argument satisfies the equation

$$\frac{dx_i}{d\lambda}\frac{dx_i}{d\lambda} + \frac{dX_i}{d\lambda}\frac{dX_i}{d\lambda} + \frac{dr_i}{d\lambda}\frac{dr_i}{d\lambda} + \frac{dR_i}{d\lambda}\frac{dR_i}{d\lambda} + \left(\frac{d\tau}{d\lambda}\right)^2 = 1.$$

the above equations can be solved with respect to derivatives in the form

$$\frac{dx_i}{d\lambda} = \frac{X_i}{Z}, \qquad \frac{dX_i}{d\lambda} = \frac{f_{x_i}}{Z}, \qquad \frac{dr_i}{d\lambda} = \frac{R_i}{Z},$$

$$\frac{dR_i}{d\lambda} = \frac{f_{r_i}}{Z}, \qquad \frac{dl_i}{d\lambda} = \frac{f_{l_i}}{Z}, \qquad \frac{d\tau}{d\lambda} = \frac{1}{Z}, \tag{4.74}$$

where $Z = (1 + X_i X_i + f_{x_i} f_{x_i} + R_i R_i + f_{r_i} f_{r_i} + f_{l_i} f_{l_i})^{1/2}$, $\quad i = \overline{2, \, n-1}$.

Let us count time from the moment $\tau = 0$ when the canopy of parachute begins filling with air. We assume that the empty part of canopy is given in the form of frustum of a cone with small angle the generator of which is continuation of the string and the filled part is a section of a sphere which is tangent to the cone (the dotted line at Fig. 4.3). This representation of the initial phase of the parachute breaking out is in good agreement with experiment [9]. Thus, the initial form of the parachute can be completely given by the radius of its inlet entrance R_c (Fig. 4.3). We assume that $R_c = 0.15R_0$ [9]. The velocity of the point of the parachute at initial instance is assumed to be equal to zero. Counting the argument λ from initial point of the problem (4.69) – (4.71), we take initial conditions in the form

$$x_i(0) = x_{i0}, \qquad X_i(0) = 0 \qquad r_i(0) = r_{i0}, \qquad R_i(0) = 0,$$

$$l_i(0) = 1, \qquad \tau(0) = 0, \qquad i = \overline{2, \, n-1}, \tag{4.75}$$

where values x_{i0}, r_{i0} correspond to the contour shown in the Fig.4.3 by dotted line.

The problem (4.74), (4.75) was integrated by means of the PC1 program for

$$R_0 = 1m, \qquad L = 1.5m, \qquad E = 2000N, \qquad m_c = 0.012\frac{kg}{m},$$

$$M = 30, \qquad p = 10^4 \frac{N}{m^2}, \qquad e = 0.015\frac{kg}{ms}, \qquad n = 26,$$

$$\frac{m_0}{m} = \begin{cases} 1.87 + 0.87(L - s), & s \geq L, \\ \\ 1, & s < L_c, \end{cases} \qquad [9].$$

The initial step of integration was equal to $h_\lambda = 0.05$ and the accuracy was equal to 10^{-4}. Configuration of the parachute for five of dimensionless times $\tau_0 = 0$, $\tau_1 = 2.5$, $\tau_2 = 5.0$, $\tau_3 = 7.5$, $\tau_4 = 10.0$ are shown at Fig. 4.3 . Variations of force along strings and bands of the canopy are shown at Fig. 4.4 for the same moments of time. The asterisk in these figures denotes the beginning of the canopy. Increasing the number of partitions along the S coordinate did not lead to substantial variations of results.

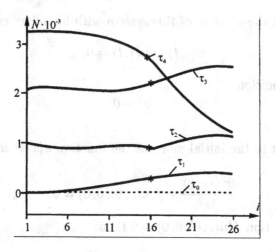

Figure 4.4.

Note, that if problem (4.74), (4.75) are solved without λ – transformation, computation process depends on the choice of the initial step of integration h_{t0}. For example, if $h_{t0} \geq 1$ the computations stoped because of overflow. Apparently the explanation is that search of the proper step size results in a large right hand sides of differential equations. λ – transformation eliminates this problem even for initial step $h_{\lambda 0} = 10$.

We consider one more model problem. The kinetic equations of neutrons in reactor can be represented for one-dimensional case in the form [53, 79]

$$\frac{\partial v}{\partial t} = av + bu + g(u, v, t),$$

$$\varepsilon \frac{\partial u}{\partial t} = \frac{\partial^2 u}{\partial x^2} + f(u, v, t). \tag{4.76}$$

Here the first equation is the kinetic equation of slow neutrons and the second one is the kinetic equation of fast neutrons; $u(x, t)$ and $v(x, t)$ are concentrations of slow and fast neutrons; $a(x, t)$, $b(x, t)$ are given functions; ε is a small parameter that is equal to the inverse value of the fast neutrons speed. We consider degenerate case when parameter $\varepsilon = 0$. Then system (4.76) can be written in the form

$$\frac{\partial v}{\partial t} = av + bu + g(u, v, t),$$

$$0 = \frac{\partial^2 u}{\partial x^2} + f(u, v, t). \tag{4.77}$$

We will find the solution of this system with boundary conditions

$$u(0, t) = u(1, t) = 0, \tag{4.78}$$

and initial condition

$$v(x, 0) = 0. \tag{4.79}$$

Clearly that in the initial moment the function $u(x, t)$ should satisfy the equation

$$\frac{\partial^2 u(x, 0)}{\partial x^2} + f(u(x, 0), 0, 0) = 0.$$

Let the function $f(u(x, 0), 0, 0) = 0$ then

$$u(x, 0) = 0. \tag{4.80}$$

To solve this problem we use the method of lines, dividing the interval $0 \le x \le 1$ into $(n-1)$ parts. After that taking into account the boundary conditions and the finite difference approximations (4.72) system (4.77) can be represented in the form

$$\frac{\partial v_i}{\partial t} = a_i v_i + b_i u_i + g_i,$$

$$\tag{4.81}$$

$$u_{j-1} - 2u_j + u_{j+1} + h f_j = 0,$$

$$i = \overline{1, n}, \qquad j = \overline{2, n-1}.$$

Here $v_i = v_i(t) = v(x_i, t)$, $u_j = u_j(t) = u(x_j, t)$, $a_i = a(x_i, t)$, $b_i = b(x_i, t)$, $g_i = g(u_i, v_i, t)$, $f_j = f(u_j, v_j, t)$, $h = 1/(n-1)$.

In derivation of the system (4.81) it has been taken into account that the boundary conditions (4.78) take the form

$$u_1 = u_n = 0. \tag{4.82}$$

This system is a system of differential – algebraic equations. Taking into account relations (4.79), (4.80) the initial conditions takes the form

$$v_i(0) = 0, \; u_j(0) = 0, \qquad i = \overline{1, n}, \qquad j = \overline{2, n-1}. \tag{4.83}$$

Let us formulate problem (4.81) – (4.83) with respect to the best argument λ. Differentiating the algebraic relations and using the meaning of the best argument, we obtain

$$v_{i,\lambda} = (a_i v_i + b_i u_i + g_i) t_{,\lambda},$$

$$u_{j-1,\lambda} - (2 - h^2 f_{j,u}) u_{j,\lambda} + u_{j+1,\lambda} +$$

$$+ h^2 f_{j,v} v_{j,\lambda} + h^2 f_{j,t} t_{,\lambda} = 0, \tag{4.84}$$

$$v_{i,\lambda} v_{i,\lambda} + u_{j,\lambda} u_{j,\lambda} + t_{,\lambda}^2 = 1.$$

To reduce this system of differential equations to the normal form we resolve it with respect to derivatives. This represents the last equation in the form

$$v_{i,\lambda}^{(k-1)} v_{i,\lambda}^{(k)} + u_{j,\lambda}^{(k-1)} u_{j,\lambda}^{(k)} + t_{,\lambda}^{(k-1)} t_{,\lambda}^{(k)} = 1,$$

where the function obtained at the previous step of integration process is marked by the superscript $(k-1)$. Thus, the used argument is close to the best argument.

Since the initial point is not a limiting point on t, the initial value of the vector $Z = (v_{1,\lambda}, v_{2,\lambda}, \ldots, v_{n,\lambda}, u_{2,\lambda}, \ldots, u_{n-1,\lambda}, t_{,\lambda})^T$ can be taken in the form (4.22). In this case we obtain exact solution of the system (4.84) if the normalization according to formulas (4.21) is performed at each step.

Obviously that dimensionality of the system (4.84) can be reduced if it is written in the form

$$u_{j-1,\lambda}^{(k)} - (2 - h^2 f_{j,u}) u_{j,\lambda}^{(k)} + u_{j+1,\lambda}^{(k)} +$$

$$+ h^2 \Big[f_{j,v}(a_j v_j + b_j u_j + g_j) + f_{j,t} \Big] t_{,\lambda}^{(k)} = 0,$$

$$u_{j,\lambda}^{(k-1)} u_{j,\lambda}^{(k)} + \tag{4.85}$$

$$+ \Big[(a_i v_i + b_i u_i + g_i)(a_i v_i + b_i u_i + g_i) + 1 \Big] t_{,\lambda}^{(k-1)} t_{,\lambda}^{(k)} = 1.$$

Here we take into account the conditions (4.82). Thus, $u_{1,\lambda} = u_{n,\lambda} = 0$.

After solving the system (4.85) with respect to derivatives and normalization according to formulas (4.21) we obtain the system of ordinary differential equations in normal form

$$\frac{du_j}{d\lambda} = u_{j,\lambda}, \qquad \frac{dv_i}{d\lambda} = (a_i v_i + b_i u_i + g_i) t_{,\lambda}, \qquad \frac{dt}{d\lambda} = t_{,\lambda}, \tag{4.86}$$

in which the quadratic norm of the right hand side is, obviously, equal to unit. This system was integrated for initial conditions

$$u_j(0) = 0 = v_i(0) = t(0) = 0, \qquad i = \overline{1, \, n}, \qquad j = \overline{2, \, n-1}, \quad (4.87)$$

and boundary conditions (4.82) by the DA1EXP program.

The following problem

$$\begin{cases} \dfrac{\partial v}{\partial t} = v + u + 4t \sin \pi x, \\[2mm] \dfrac{\partial^2 u}{\partial x^2} + \pi^2 v = 0. \end{cases} \qquad (4.88)$$

$$u(0, t) = u(1, t) = 0, \qquad (4.89)$$

$$v(x, 0) = 0. \qquad (4.90)$$

was considered as an example.

If the the following form

$$v(x, t) = \sum_{i=1}^{\infty} V_i(t) \sin i\pi x,$$

$$u(x, t) = \sum_{i=1}^{\infty} U_i(t) \sin i\pi x,$$

of solution of the problem (4.88) – (4.90) is assumed then it is easy to show that analytical solution is

$$v(x, t) = u(x, t) = \left(e^{2t} - 1 - 2t \right) \sin \pi x. \qquad (4.91)$$

The numerical solution obtained for $t \in [0, 1]$ was compared with the analytical solution (4.91) for $t = 1$.

If the initial step is decreased the accuracy of computation increases. The numerically calculated functions $v(x, t)$, $u(x, t)$ are symmetric with respect to the middle of the interval $0 \leq x \leq 1$. Increase of the number of dividing points n (we consider $n = 11$ and $n = 16$) reduces the difference between functions $v(x, t)$ and $u(x, t)$.

The errors Δ_v and Δ_u were computed as difference between numerical and analytical values of functions v and u. For initial step $h_{\lambda 0} = 0.05$, $n = 11$ and at the moment $t = 1$ the errors were equal to $\Delta_v = 0.163$ and $\Delta_u = 0.202$. For $h_{\lambda 0} = 0.025$ they were equal to $\Delta_v = 0.091$ and

$\Delta_u = 0.128$. For $h_{\lambda 0} = 0.0125$ they were equal to $\Delta_v = 0.0386$ and $\Delta_u = 0.0755$. For $n = 16$ and $h_{\lambda 0} = 0.025$ the errors were equal to $\Delta_v = 0.075$ and $\Delta_u = 0.094$.

As it was expected, larger error was accumulated in the variable that was not differentiated (i.e. in the function u).

Chapter 5

FUNCTIONAL - DIFFERENTIAL EQUATIONS

In this chapter we consider one more class of problems the solution of which can be obtained by the parametric continuation method. This is the initial value problem (the Cauchy problem) for the functional differential equations. The equations with nonlocal retarded argument and integro differential equations can be included into this class of problems.

1. The Cauchy problem for equations with retarded argument

The problems of this kind appear in a variety of applied problems such as spreading of epidemics, dynamics of populations, modeling of the cordial – pulmonary systems, theory of control, etc.

Sufficiently detailed review of methods for solution of the equations with retarded argument are given in [81, 31, 20]. Apparently, the application of Euler method for numerical integrating of such problems was qualitatively considered in [29, 30]. The first computation program, which uses Euler method and piecewise constant or piecewise linear approximations is given in [36].

The analysis shows that programs developed for solution of ordinary differential equations can be used for numerical integration of the equations with retarded argument. In [122] the comparison of some most popular numerical programs for solution of the Cauchy problem for stiff and unstiff equations with retarded argument is given .

Note that application of methods, which order is greater than one, to the retarded equation is complicated. This is because only the solutions, which are differentiable sufficient number of times, are well approximated by these methods. Thus, these methods give good results only on the intervals of argument where the solution is sufficiently smooth.

Let us consider the Cauchy problem for a system of ordinary differential equations with retarded argument $\tau_{ij} \geq 0$

$$\frac{dy_i}{dt} = f_i\Big(t,\ y_1(t),\ \ldots\ ,y_n(t),\ y_1[t - \tau_{1i}(t)],\ \ldots\ ,$$

$$,\ \ldots\ ,\ y_n[t - \tau_{ni}(t)]\Big),$$

$$y_i = \varphi_i(t), \qquad \min(t_0, t_{*i}) \leq t \leq t_0, \qquad\qquad (5.1)$$

$$t_{*i} = \min_{\substack{1 \leq j \leq n \\ t \in [t_0, T]}} [t - \tau_{ij}(t)], \qquad i = \overline{1,\ n}.$$

Here $\varphi_i(t)$ are given functions. This problem is to be solved on the interval $[t_0, T] \ni t$.

Suppose that the solution of problem (5.1) is given by the relation

$$F_j(y_1,\ \ldots\ ,y_n,\ t) = 0, \qquad j = \overline{1,\ n}, \qquad\qquad (5.2)$$

which determines the integral curve in $(n + 1)$ - dimensional Euclidean space $^{n+1}$.

Obtaining this curve can be considered as a solution of the system of equations (5.2) containing the argument t as a parameter. We will solve this system by the parametric continuation method. As we have already seen, in this case the parameter t is not always the best (or even satisfactory) parameter of continuation. Thus, the problem of finding the best continuation parameter can be stated for solution (5.2) and, hence, for the problem (5.1). As in previous chapter, this problem will be solved locally, i.e., in a small neighborhood of each point of the integral curve where continuation is performed. In this neighborhood we will assume that all y_i and t are the functions of the argument μ given by the relation

$$d\mu = \alpha_i dy_i + \alpha_{n+1} dt, \qquad i = \overline{1,\ n}. \qquad\qquad (5.3)$$

Here α_k $(k = \overline{1,\ n+1})$ are the components of the unit vector $\alpha = (\alpha_1,\ \ldots\ ,\alpha_{n+1})^T$ specifying the direction in which the argument μ is counted.

The equations of the solution continuation we obtain by differentiating relations (5.2) as a composite functions with respect to μ

$$F_{j,y_i}\, y_{i,\mu} + F_{j,t}\, t_{,\mu} = 0, \qquad i, j = \overline{1,\ n}, \qquad\qquad (5.4)$$

and dividing the equation (5.3) by $d\mu$

$$\alpha_i y_{i,\mu} + \alpha_{n+1} t_{,\mu} = 1, \qquad i = \overline{1, n}. \qquad (5.5)$$

If the Jacobi matrix (F_{i,y_j}) is singular the relations (5.4) can be written in the form

$$y_{i,\mu} = -(F_{j,y_i})^{-1} F_{j,t} t_{,\mu}.$$

On the base of the theorem about differentiation of the implicit functions [40] we obtain

$$-(F_{j,y_i})^{-1} F_{j,t} = \frac{dy_i}{dt}.$$

Thus, taking into account equations (5.1), the relations (5.4), (5.5) reduce to the form

$$
\begin{aligned}
\alpha_i y_{i,\mu} + \alpha_{n+1} t_{,\mu} &= 1, \\
y_{i,\mu} - f_i t_{,\mu} &= 0, \qquad i = \overline{1, n}.
\end{aligned}
\qquad (5.6)
$$

This system is a system of the continuation equations for obtaining the curve of the solution set given by the relations (5.2). Apparently, this curve is at the same time the integral curve of problem (5.1).

To find this curve we must represent the system (5.6) in normal Cauchy form, i.e.,we must resolve this system with respect to derivatives. The success of this operation depends on the conditionality of the system. The conditionality, in turn, depends on the choice the parameter – argument μ determined by the vector $\alpha = (\alpha_1, \ldots, \alpha_{n+1})^T$. Structure of the system (5.6) is the same as the structure of system (1.37) that was considered in chapter 1 where the theorem about the best continuation parameter for a system of nonlinear equations was proved. Thus, the best parameter λ will be realized if the vector α takes the value $\alpha = (y_{1,\lambda}, \ldots, y_{n,\lambda}, t_{,\lambda})^T$. Then system (5.6) can be solved analytically with respect to derivatives. Since the argument λ is not contained explicitly in the equations, we will count it from initial point of the problem (5.1). As a result, we obtain the following Cauchy problem

$$\frac{dy_i}{d\lambda} = f_i\left(t(\lambda),\ y_1(\lambda),\ \ldots,\ y_n(\lambda),\ y_1\left(t(\lambda) - \tau_{1i}(t(\lambda))\right),\ \ldots,\right.$$

$$,\ \ldots,\ y_n\left(t(\lambda) - \tau_{ni}(t(\lambda))\right)\left(1 + f_j f_j\right)^{-\frac{1}{2}},$$

$$\frac{dt}{d\lambda} = \left(1 + f_j f_j\right)^{-\frac{1}{2}}, \qquad i,j = \overline{1,\ n},$$

$$y_i(0) = \varphi_i(t_0), \qquad t(0) = t_0,$$

$$y_i = \varphi_i(t), \qquad \min(t_0, t_{*i}) \le t \le t_0,$$

$$t_{*i} = \min_{1 \le j \le n}\left\{t(\lambda) - \tau_{ij}(t(\lambda))\right\},$$

$$\lambda \in [0, L], \qquad T = t(L), \qquad i = \overline{1,\ n}.$$

(5.7)

Thus, the following theorem is proved.

Theorem 5.1. In order to formulate the Cauchy problem for a system of ordinary differential equations with retarded argument (5.1) with respect to the best argument, it is necessary and sufficient to choose the arc length λ measured along the integral curve of this problem as this parameter. Then the transformed system takes the form (5.7).

As a model example, that shows the advantages of the suggested transformation, we consider numerical solution of the Cauchy problem for the equation with retarded argument

$$\frac{dy}{dt} = \frac{1}{3y^2(t - 2\tau)}, \qquad y = t^{\frac{1}{3}}, \qquad -\tau \le t \le \tau > 0. \qquad (5.8)$$

We find the solution of this problem within the interval $\tau \le t \le 3\tau$. By applying the method of steps [31], solution of the problem (5.8) is reduced to the Cauchy problem without delay

$$\frac{dy}{dt} = \frac{1}{3(t - 2\tau)^{\frac{2}{3}}}, \qquad y(\tau) = y_\tau = \tau^{\frac{1}{3}}. \qquad (5.9)$$

Particular integral of the problem is of the form

$$\left(y - y_\tau - \tau^{\frac{1}{3}}\right)^3 - t + 2\tau = 0. \qquad (5.10)$$

Note that numerical solution of the problem (5.8) is complicated by the fact that the right hand sade of the equation becomes infinite in

the point $t = 2\tau$ of the integration interval. A FORTRAN – program
was developed to solve this problem. The explicit Euler method was
used for integrating the equation. The calculated nodal values of chosen
function were stored in memory. The piecewise linear interpolation were
used to find the internodal values of function y. An attempt to find
the solution of the problem (5.8) for $\tau = 1$ by means of this program
was unsuccessful. The computational error which was estimated by the
expression

$$\Delta = \left(y^* - y_\tau - \tau^{\frac{1}{3}}\right)^3 - t + 2\tau,$$

where y^* is a numerical solution of the problem, took catastrofical value
at $t = 2\tau = 2$.

After λ – transformation, the problem (5.8) is reduced to the Cauchy
problem of the form

$$\frac{dy}{d\lambda} = \frac{1}{\sqrt{1+\omega^2}}, \qquad \frac{dt}{d\lambda} = \frac{\omega}{\sqrt{1+\omega^2}},$$

$$y(0) = y_\tau = \tau^{\frac{1}{3}}, \quad t(0) = \tau, \tag{5.11}$$

$$y = t^{\frac{1}{3}}, \quad -\tau \le t \le \tau, \quad \omega = 3y^2(t - 2\tau).$$

The integration of the problem for $t \in [\tau, 3\tau]$ was successful.

To analyze the behavior of the numerical solution when the right hand
side of equation goes to infinity, we formulate the Cauchy problem (5.8)
in Cartesian coordinate system (z, x) rotated at angle α (Fig. 5.1) with
respect to coordinate system (y, t).

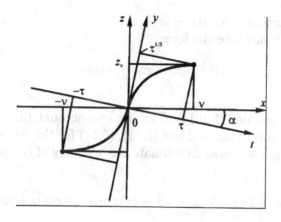

Figure 5.1.

Relationship between these two coordinate systems is given by the formulas

$$\begin{bmatrix} z \\ x \end{bmatrix} = R(\alpha) \begin{bmatrix} y \\ t \end{bmatrix}, \qquad \begin{bmatrix} y \\ t \end{bmatrix} = R^{-1}(\alpha) \begin{bmatrix} z \\ x \end{bmatrix}, \tag{5.12}$$

$$R(\alpha) = \begin{bmatrix} \cos \alpha & -\sin \alpha \\ \sin \alpha & \cos \alpha \end{bmatrix} - .$$

For $\tau = 1$ the problem (5.8) is transformed to the form

$$\frac{dz}{dx} = \frac{\cos \alpha - 3Q \sin \alpha}{\sin \alpha + 3Q \cos \alpha}, \tag{5.13}$$

where $Q = [z(x - 2\nu) \cos \alpha + (x - 2\nu) \sin \alpha]^2$,

$$z \cos \alpha + x \sin \alpha = (-z \sin \alpha + x \cos \alpha)^{\frac{1}{3}}, \tag{5.14}$$
$$-\nu \leq x \leq \nu.$$

The projections of point $t = \tau = 1$, $y(\tau) = \tau^{\frac{1}{3}} = 1$ onto axes x and z are equal to

$$\nu = \sin \alpha + \cos \alpha, \qquad z(\nu) = z_\nu = \cos \alpha - \sin \alpha.$$

Particular integral of the problem on the interval $\nu \leq x \leq 3\nu$ can be obtained analytically by means of the steps method if problem (5.9) for $\tau = 1$ is transformed according to (5.12) to the form

$$\frac{dz}{dx} = \frac{\cos \alpha - 3P \sin \alpha}{\sin \alpha + 3P \cos \alpha}, \tag{5.15}$$

where $P = [-z \sin \alpha + x \cos \alpha - 2\nu]^{\frac{2}{3}}$.

Initial conditions take the form

$$z(\nu) = z_\nu = \cos \alpha - \sin \alpha. \tag{5.16}$$

Writing expressions (5.10) and taking into account (5.12) we obtain particular integral of the problem (5.13), (5.14) for the interval $\nu \leq x \leq 3\nu$. This relation was used to estimate the accuracy of computations by the formula

$$\Delta = z^* \cos \alpha + x \sin \alpha - z_\nu - \nu^{\frac{1}{3}} + (z^* \sin \alpha - x \cos \alpha + 2\nu)^{\frac{1}{3}}, \tag{5.17}$$

where z^* is the numerical solution of the problem.

The behavior of common logarithm of modulus of error Δ computed by the formula (5.17) is shown at Fig. 5.2 as a function of the angle of coordinate system rotation α. The value of error Δ at the end of integration interval (i.e. for $t = 3\nu$) is demonstrated . The dotted line corresponds to numerical solution of the problem (5.13), (5.14). Solid line corresponds to the same problem after λ – transformation.

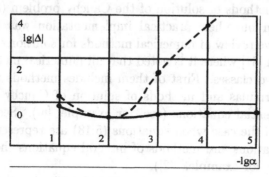

Figure 5.2.

Initial value of the function $z(x)$, $-\nu \leq x \leq \nu$ was determined from equation (5.14) which was solved numerically by Newton - Raphson method with the accuracy 10^{-4}.

Problems were integrated with the same steps of the variables x and λ equal to 0.1.

It can be seen from Fig. 5.2 that the error of the λ – transformed problem does not depend on the angle α, whereas the error of traditional approach increases indefinitely when α tends to zero.

2. The Cauchy problem for Volterra's integro – differential equations

We consider numerical solution of the Cauchy problem for system of equations of the form

$$F_i\left(t, y(t), \dot{y}(t), \int_{t_0}^{t} K_i[t, \xi, y(\xi)]d\xi\right) = 0, \qquad (5.18)$$

for initial conditions

$$y(t_0) = y_0 \qquad (5.19)$$

$$y: {}^1 \to {}^n, \qquad t \in {}^1, \qquad \dot{y} = \frac{dy}{dt}, \qquad i = \overline{1, n}$$

Reviews of results obtained for such problems are given in [81, 20, 5].

Equations (5.18) are called equations with infinite retardation since, in general case, the derivative $\dot{y}(t)$ depends on all previous values of the function $y(t)$. Like the equations with retarded argument such equations are an example of Volterra's functional – differential equations.

Problems of the type (5.18), (5.19) are practically important. In particular they are used for description of competing populations.

Numerical methods of solution of the Cauchy problem (5.18), (5.19) are well known but their practical implementation was limited [37]. Rather extensive review of numerical methods for solution of such problems is given in [81] where it is noted that all numerical methods can be divided into two classes. First of them includes methods which adapt directly the formulas and methods of solution of Cauchy problem for ordinary differential equations (see, for example, [6]). Second class is associated with the case when equations (5.18) are represented in form of integral equations and methods of integral equations theory are implemented (see, for example, [62]).

Here problem (5.18), (5.19) is formulated with respect to the best argument. The algorithm considered above will be implemented.

Note that formal approach to solution of this problem was considered in [107].

Integral of the problem (5.18), (5.19)

$$f(y,t) = 0, \qquad f : {}^{n+1} \to {}^{n} \tag{5.20}$$

defines a curve in $(n+1)$ - dimensional Euclidean space ${}^{n+1}$. The construction of this curve can be represented as a process of solution continuation with respect to parameter t. Such an approach allows to formulate a problem of finding the best continuation parameter, i.e., the best argument.

Let the values y and t to be functions of some argument μ which locally, i.e., in small neighborhood of each point of the problem integral curve is represented by the relation

$$d\mu = \alpha_i dy_i + \alpha_{n+1} dt, \qquad i = \overline{1,\,n}, \tag{5.21}$$

where α_j $(j = \overline{1,\,n+1})$ are components of the unit vector $\alpha = (\alpha_1, \ldots, \alpha_{n+1})^T$ that defines the direction along which the parameter μ is counted.

We obtain equations of solution continuation for the problem (5.20) if relation (5.21) is divided by $d\mu$ and expression (5.20) is differentiated with respect to μ

$$\begin{aligned} \alpha_i y_{i,\mu} + \alpha_{n+1} t_{,\mu} &= 1, \\ f_{,y_i} y_{i,\mu} + f_{,t} t_{,\mu} &= 0. \end{aligned} \tag{5.22}$$

Here $y_{i,\mu} = \dfrac{dy_i}{d\mu}, \qquad t_{,\mu} = \dfrac{dt}{d\mu}, \qquad f_{,y_i} = \dfrac{\partial f}{\partial y_i}, \qquad f_{,t} = \dfrac{\partial f}{\partial t}.$

But this approach is not constructive because the integral (5.20) is not known.

Continuation equations can be obtained in a different way. Let us linearize the vector function $F = (F_1, \dots, F_n)^T$ at the neighborhood of certain values $y_i = y_i^*$. Then we obtain

$$F^* + F_{,\dot{y}_i}^* \, (\dot{y}_i - \dot{y}_i^*) = 0, \qquad i = \overline{1, \, n}.$$

Here functions F^* and $F_{,\dot{y}_i}^*$ are calculated for $\dot{y}_i = \dot{y}_i^*$.
Taking into account the relations

$$\dot{y}_i = \frac{y_{i,\mu}}{t_{,\mu}}, \qquad \dot{y}_i^* = \frac{y_{i,\mu}^*}{t_{,\mu}^*},$$

we obtain the following form of continuation equations

$$\begin{aligned} &\alpha_i y_{i,\mu} + \alpha_{n+1} t_{,\mu} = 1, \\ &t_{,\mu}^* \, F_{,\dot{y}_i}^* \, y_{i,\mu} + \left(F^* t_{,\mu}^* - F_{,\dot{y}_i}^* \, y_{i,\mu}^* \right) t_{,\mu} = 0. \end{aligned} \qquad (5.23)$$

Integral curve of the problem (5.18), (5.19) can be constructed as a result of integration of differential equations obtained after solving of the continuation equations (5.23) with respect to derivatives. The conditionality of the system depends on choice of the argument μ determined by vector α. The structure of system (5.23) is fully consistent with the structure of system (1.37) considered earlier when the best continuation parameter for a system of nonlinear algebraic or transcendental equations with a parameter was chosen. A parameter, which ensures the best conditionality of the continuation equations system, is the arc length λ counted along the curve of solution set. In the considered case the curve is the integral curve of problem (5.18), (5.19) and the parameter λ is its best argument.

In according to Cramer's rule the solution of system (5.23) can be represented in the form

$$\frac{dy_i}{d\lambda} = (-1)^{i+1} \frac{\Delta_i}{\Delta}, \qquad \frac{dt}{d\lambda} = (-1)^{n+2} \frac{\Delta_{n+1}}{\Delta}, \qquad i = \overline{1, \, n}, \qquad (5.24)$$

where Δ is the determinant of system, Δ_i is the determinant obtained by deleting the last row and i - th column of Δ. These determinants satisfy the relation

$$\Delta^2 = \Delta_j \Delta_j, \qquad j = \overline{1, \, n+1}, \qquad (5.25)$$

which shows that squared norm of the right hand side of system of equation (5.24) are always equal to one. If argument λ is counted from the initial point of problem (5.18), (5.19) the initial conditions take the form

$$y(0) = y_0, \qquad t(0) = t_0. \tag{5.26}$$

Thus, on the base of the results of chapter 1 the following theorem is proved.

Theorem 5.2. In order to formulate the Cauchy problem for a system of integro – differential equations (5.18), (5.19) with respect to the best argument it is necessary and sufficiently to choose the arc length λ measured along the integral curve of the problem as such a parameter. Then the problem (5.18), (5.19) is formulated in the form (5.24), (5.26) and the right hand sides of system (5.24) satisfy the equation (5.25).

As before, transformation of the problem (5.18), (5.19) to the problem (5.24), (5.26) will be named λ – transformation. We demonstrate one of its advantages on the example of the following Cauchy problem

$$\frac{dy}{dt} = \frac{1}{3y^2} + S(t, y),$$
$$y(-1) = -1, \tag{5.27}$$

where $S(t, y) = \int\limits_{-1}^{t} \xi^{\frac{2}{3}} y(\xi)\, d\xi + \frac{1}{2}\left(1 - t^2\right).$

This problem has exact solution $y(t) = t^{1/3}$ for which the right hand sides of equation (5.27) becomes infinite at $t = 0$. After application of the λ – transformation the problem takes the form

$$\frac{dy}{d\lambda} = \frac{\varphi}{z}, \qquad \frac{dt}{d\lambda} = \frac{y^2}{z},$$
$$y(0) = -1, \quad t(0) = -1. \tag{5.28}$$

Here $\varphi = \frac{1}{3} + y^2 S(t, y), \qquad z = \left(y^4 + \varphi^2\right)^{\frac{1}{2}}.$

Problems (5.27) and (5.28) were integrated numerically within the interval $t \in [-1, 1]$ by the Euler method. The integral contained in the right hand sides of equations was evaluated by trapezoid formula with variable step. Problem (5.27) was integrated with the step $h_t = 0.002$. The error Δ was estimated by the formula

$$\Delta = y^* - t^{\frac{1}{3}}, \tag{5.29}$$

where y^* is numerical solution of the problem. This error increased abruptly by two orders of magnitude at $t = 0$ (from $\Delta = -0.175$ to $\Delta = 17.4$). For λ – transformed problem integrated with the step $h_\lambda = 0.02$ no sharp increase of error was observed.

Let us consider the equation containing more complicated integral, which has, however, the same solution. Let function $S(t, y)$ in the problem (5.27) to have the form

$$S(t, y) = \int\limits_{-1}^{t} \xi^{2t} y(\xi) \, d\xi + \delta(t),$$

$$\text{where } \delta(t) = \begin{cases} \dfrac{1}{2\left(t + \dfrac{2}{3}\right)} \left(1 - t^{2\left(t + \frac{2}{3}\right)}\right), & t \neq -\dfrac{2}{3}, \\ \ln \dfrac{2}{3}, & t = -\dfrac{2}{3}. \end{cases}$$

When this problem was solved numerically by means of the algorithm presented above, the error (5.29) was increased from $\Delta = -0.138$ to $\Delta = 0.226 \cdot 10^3$ after passing the value of argument $t = 0$. (Integration step was $h_t = 0.002$.) After applying λ – transformation to the problem no sharp increase of error was observed.

Chapter 6

THE PARAMETRIC APPROXIMATION

The problem of approximation plays an extremely important role in modern computation mathematics because many numerical methods are based on the concept of approximation. The parametric approximation of plane curves involving the choice of the best parameter is examined in this chapter. It is shown, that such approach has certain advantages over the usual methods.

The advantages of parametric approximation in the case when a complicated curves are approximated by parametrix splines are discussed in detail in [125]. It is important to note, though, that in practice only ordered arrays of points on the curves to be approximated are given , and no information on the method of parametrization needed to construct the splines is given. In that situation, natural parametrization is used to construct parametric splines in [125], and the length of the arc of the curve from the initial point is taken as the parameter.

Cubic and parametric splines are used to model curves in [90]. The problem of defining the approximating spline is formulated either as the problem of reconstructing an object, or as a problem of approximation with given accuracy. In the first case, an interpolating or smoothing spline is sought for a given set of points, while in the second case a set of points is chosen for a given curve so that the interpolating spline approximates the curve with given accuracy. The parameter is chosen as the length of the broken line or length of the arc of the curve, which is found by iteration formulas.

The parametric approximation of closed curve in the plane by polynomial splines is considered in [104]. Estimates of the approximation

149

error is given in Hausdorff metrics. The value

$$r(E,\ F) = \max[\max_{P \in E} \min_{Q \in F} \rho(P,\ Q),\quad \max_{P \in F} \min_{Q \in E} \rho(P,\ Q)],$$

is regarded as Hausdorff's metric of distance between two bounded closed sets of points E and F. Here $\rho(P,\ Q)$ is Euclidean distance between points P and Q.

Exact estimates of the approximation error of curves by parametric Hermitian splines in Euclidean and Hausdorff metrics are obtained in [87, 118]. The length of the broken line or curve is used as the parameter.

It should be noted that the problem of choosing of the best parameter was not discussed in any of this studies.

1.   The parametric interpolation

Unlike conventional interpolation (see [7], for example) we will formulate the problem of parametric interpolation as a problem of constructing two functions $X(\mu)$ and $Y(\mu)$ which, for certain selected values of the parameter μ_i, $i = \overline{1,\ n}$, take pre-assigned values x_i, y_i, $i = \overline{1,\ n}$, that is,

$$X(\mu_i) = x_i, \qquad Y(\mu_i) = y_i, \qquad i = \overline{1,\ n}. \tag{6.1}$$

Obviously, in this formulation the choice of the quantities μ_i is not unique. The numbers μ_i only need to form a monotone sequence. For example, μ_i could be chosen as the indices of the number pairs x_i, y_i, that is, we could put $\mu_i = i$.

Generally speaking, the functions $X(\mu)$ and $Y(\mu)$ can be functions of different classes or different functions of the same class. For example, from the standpoint of parametric interpolation, the conventional problem of constructing an interpolation polynomial $y = L(x)$ is formulated as the problem of constructing a polynomial $y = L(\mu)$ for $x = \mu$.

The fact that the choice of the parameter μ is not unique allows to raise the question of choosing the best (at least in some sense) parameter. The importance of the task is clear from the results obtained in [32], where it was established in a study of a parametric cubic spline that the choice of parameter has a considerable influence on the interpolation curve, which might even contain loops in some cases.

If two such functions:

$$x = X(\mu) \qquad y = Y(\mu), \tag{6.2}$$

which satisfy conditions (6.1) have been constructed, the problem of computing inter-nodal values can be formulated in two ways.

1. Given μ, it is required to compute x and y. The solution of this problem is trivial and reduces to computing $x = X(\mu)$ and $y = Y(\mu)$.

2. Given x (or y),it is required to compute y (or x).

The latter formulation is typical for standard interpolation. For parametric interpolation, μ must first be found from given x by solving the nonlinear equation $x - X(\mu) = 0$ and the obtained μ is used to compute y as $y = Y(\mu)$. Thus, in the most general case with parametric interpolation, computing inter-nodal values reduces to solving a system of two nonlinear equations

$$x - X(\mu) = 0, \qquad y - Y(\mu) = 0. \tag{6.3}$$

We will eliminate μ from these equations. The conditions on the functions $X(\mu)$ and $Y(\mu)$ for which this can be done will not be discussed here, but are clearly not too burdensome. As a result, the problem of calculating inter-nodal values of x and y will reduce to solving the nonlinear equation

$$F(x, \, y) = 0, \tag{6.4}$$

and, by (6.1) and (6.2), the n solutions x_i, y_i of this equations are known, i.e.,

$$F(x_i, \, y_i) = 0, \qquad i = \overline{1, \, n}. \tag{6.5}$$

When the interpolation problem is formulated in this way, it can be considered from the point of view of the method of continuation of the solution with respect to parameter, and the parameter μ can be introduced as the continuation parameter, that is, the solutions x and y of equation (6.4) can be sought as functions of a parameter μ:

$$x = x(\mu), \qquad y = y(\mu). \tag{6.6}$$

It was proved in chapter 1 that the system of linearized equations corresponding to equation (6.4) obtained at each step of continuation of the solution will be best-conditioned if and only if, the continuation parameter μ is chosen as the arc length λ of the curve of the set of solutions of equation (6.4), i.e., the curve $y(x)$ or $x(y)$ which corresponds to this solution. The differential of the arc of the curve $d\lambda$ will satisfy the equation

$$(d\lambda)^2 = (dx)^2 + (dy)^2. \tag{6.7}$$

Later on the parameter providing the best conditionality to the linearized system of continuation equations will be named as the best parameter.

Thus, the following theorem have been proved.

Theorem 6. In order to formulate the problem of parametric interpolation of the curve with respect to the best parameter, it is necessary and sufficient to choose the arc length of the interpolated curve as this parameter.

Given equation (6.4) satisfying conditions (6.5), the interpolation functions (6.6) $x(\lambda)$ and $y(\lambda)$ can be obtained as solutions of the continuation equation (6.4) with respect to λ:

$$F_{,x} \frac{dx}{d\lambda} + F_{,y} \frac{dy}{d\lambda} = 0,$$

$$F_{,x} = \frac{\partial F}{\partial x}, \qquad F_{,y} = \frac{\partial F}{\partial y}.$$

This equation, together with equality (6.7), can be solved for $dx/d\lambda$, $dy/d\lambda$ and the problem of constructing interpolating functions can thus be reduced to interpolating the following Cauchy problem:

$$\frac{dy}{d\lambda} = -\frac{F_{,x}}{\sqrt{F_{,x}^2 + F_{,y}^2}}, \qquad y(0) = y_1,$$

$$\frac{dx}{d\lambda} = \frac{F_{,y}}{\sqrt{F_{,x}^2 + F_{,y}^2}}, \qquad x(0) = x_1. \tag{6.8}$$

The parameter λ here is taken from the first interpolation node (x_1, y_1). Some advantages of such a formulation of the Cauchy problem have been noted in chapter 2.

The conventional problem of constructing an interpolation function $y = f(x)$ of a given class, an interpolation polynomial for instance, can be solved in this form merely by taking equation (6.4) as

$$F(x, y) = y - f(x) = 0. \tag{6.9}$$

Then, by virtue of (6.9), $F_{,y} = 1$, $F_{,x} = -df/dx$, the second of the continuation equations (6.8) can be simplified and the nodal values of the parameter can be found with the recurrence formula

$$\lambda_1 = 0, \qquad \lambda_{i+1} = \lambda_i + \int_{x_i}^{x_{i+1}} \sqrt{1 + \left(\frac{df}{dx}\right)^2}\, dx, \qquad i = \overline{1, n-1}.$$

The symmetry with respect to x and y of the parametric interpolation (6.3) which uses the best parameter λ is also interesting. In order to construct interpolating function $X(\lambda)$, $Y(\lambda)$, the nodal values λ_i must be assigned. But these, in turn, cannot be computed before the functions $X(\lambda)$, $Y(\lambda)$ have be assigned. Without discussing the general formulation of the problem here we note, that, to a first approximation, the nodal values $\lambda_i^{(1)}$ $(i = \overline{1, n})$ can be taken as the length of the broken line joining the points (x_i, y_i) $(i = \overline{1, n})$ in the x, y plane, starting from point x_1, y_1, that is, by the recurrence formula

$$\lambda_1^{(1)} = 0, \qquad \lambda_{i+1}^{(1)} = \lambda_i^{(1)} + \Delta\lambda_i,$$

$$\Delta\lambda_i = \sqrt{(x_{i+1} - x_i)^2 + (y_{i+1} - y_i)^2}, \qquad (6.10)$$

$$i = \overline{1, n-1}.$$

Once interpolation functions of the first approximation $X^{(1)}(\mu)$, $Y^{(1)}(\mu)$ such that

$$X^{(1)}\Big|_{\mu=\lambda_i^{(1)}} = x_i, \qquad Y^{(1)}\Big|_{\mu=\lambda_i^{(1)}} = y_i,$$

have been constructed, the accuracy of the nodal values λ_i can be improved:

$$\lambda_1^{(2)} = 0, \qquad \lambda_{i+1}^{(2)} = \lambda_i^{(2)} + \int_{\lambda_i^{(1)}}^{\lambda_{i+1}^{(1)}} \sqrt{\left(\frac{dX^{(1)}}{d\mu}\right)^2 + \left(\frac{dY^{(1)}}{d\mu}\right)^2}\, d\mu,$$

$$i = \overline{1, n-1},$$

and interpolation function of the second approximation $X^{(2)}(\mu)$, $Y^{(2)}(\mu)$ constructed, and so on.

Without discussing the convergence of this process here, we merely note that for practical purposes of parametric interpolation, a first approximation is quite sufficient. The length of a broken line of the form (6.10) has been used as a parameter in [125, 90, 32] when constructing a parametric cubic spline. The fact that the length of the broken polygon is close to the best continuation parameter ensures that the obtained solution of problems of determining inter-nodal values of x (or y) form given y (or x) is nearly best-conditioned .

Note also that the best continuation parameter considered in chapter 1 ensures minimal quadratic error of the solution of the corresponding linearized equation for (6.4) when the arguments are given with errors.

As a consequence, when the length of the broken line (6.10) is chosen as a parameter, the variations of the interpolating functions $X(\mu)$, $Y(\mu)$ arising from errors at the interpolation nodes will be close to minimal.

In addition, parametric interpolation with that parameter enables many-valued functions $y(x)$ and $x(y)$ to be interpolated, including functions whose graphs are closed curves or curves which form loops. The examples considered below show that curves constructed by means of parametric interpolation of this kind are closer to the graph of the given function.

As our first example, we will consider the approximation by polynomials of the unit semi-circle $x^2 + y^2 = 1$, $y \geq 0$, with interpolation nodes (x_i, y_i) $(i = \overline{1, n})$, spaced evenly along its circumference. The ordinary interpolation problem is solved using the Lagrange polynomial

$$y(x) = \sum_{i=1}^{n} y_i \prod_{j=1 j \neq i}^{n} \frac{x - x_j}{x_i - x_j}. \tag{6.11}$$

The parametric interpolation is realized in the form of two Lagrange polynomials in terms of the length of the broken line (6.10):

$$x(\lambda) = \sum_{i=1}^{n} x_i \prod_{j=1 j \neq i}^{n} \frac{\lambda - \lambda_j}{\lambda_i - \lambda_j},$$

$$\tag{6.12}$$

$$y(\lambda) = \sum_{i=1}^{n} y_i \prod_{j=1 j \neq i}^{n} \frac{\lambda - \lambda_j}{\lambda_i - \lambda_j}.$$

The interpolation error δ is computed from the formula

$$\delta_k = 1 - x^2 - y^2, \qquad k = 1, \ 2,$$

where the number $k = 1$ corresponds to ordinary interpolation, defined by polynomial (6.11), and the number $k = 2$ corresponds to parametric interpolation, using polynomials (6.12).

We denote the largest value of the absolute value of δ_k by Δ_k.

Fig. 6.1 shows the dependence of the logarithm of Δ on the number of interpolation nodes n. The dashed curve represents $\lg \Delta_1$, and continuous curve represents $\lg \Delta_2$. This is seen more clearly in Fig. 6.2, where the values of δ_1 and δ_2 (the dashed and continuous curves, respectively) are represented as functions of the arc length λ in the case of seven interpolation nodes ($n = 7$). The error Δ_1 is obviously more than two orders larger than the error of parametric interpolation Δ_2 and when $n = 9$, for example, the difference is even greater, reaching almost three orders of magnitude.

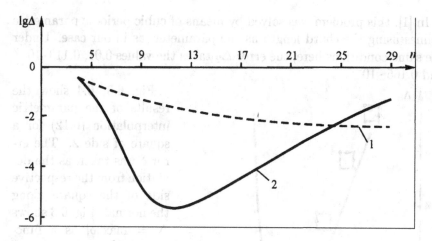

Figure 6.1.

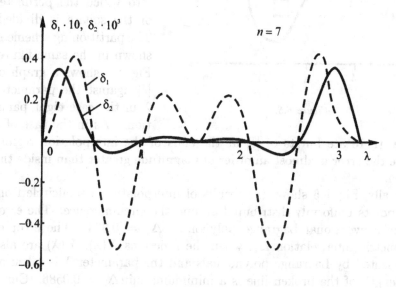

Figure 6.2.

Note that the error δ_2 has bursts at the ends of the interpolation region, whereas it is almost an order of magnitude smaller inside that region. It is also worth noting that the ordinary interpolation error Δ_1 in this example is as small as possible when the interpolation nodes coincide with the zeros of the Chebyshev polynomials (see, [7]). For $n = 7$, for example, both Δ_1 and Δ_2 are more than three times smaller.

In the case of parametric interpolation of the whole unit circle with nodes arranged uniformly along the circumference with 4, 8 and 12 partitions, the errors Δ_2 are: 0.39, $0.135 \cdot 10^{-1}$, $0.88 \cdot 10^{-5}$.

In [1], this problem was solved by means of cubic periodic parametric splines using the chord length as the parameter, as in our case. Under the same conditions here, the error Δ_2 takes the values 0.01, $0.113 \cdot 10^{-2}$ and $0.165 \cdot 10^{-3}$.

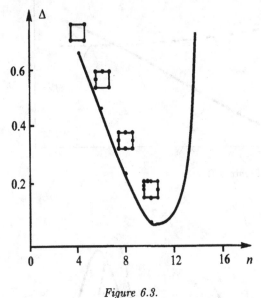

Figure 6.3.

Fig. 6.3, 6.4 show the results of the parametric interpolation (6.12) for a square of side 2. The error δ was taken as the deviation from the respective side of the square along the normal. Fig. 6.3 shows $\Delta = \max|\delta|$ as a function of the number of parts into which the perimeter of the square is divided. The partitioning scheme is shown in the same figure. Fig. 6.4 shows a graph of $|\delta|$ against the parameter λ in the case eight partitions. As in the case of a circle, there are bursts of $|\delta|$ at the ends of the interpolation region, where the error is almost an order of magnitude greater than inside the region.

Finally, Fig. 6.5 shows the results of interpolating a semicircle from four points uniformly distributed around the circumference. The error of the conventional Lagrange polynomial $\Delta_1 = 0.333$. The error of parametric interpolation Δ_2, when the functions $X(\lambda)$, $Y(\lambda)$ are also represented by Lagrange polynomials and the parameter λ is taken as the length of the broken line is a minimum: $\min \Delta_2 = 0.0586$. Curve

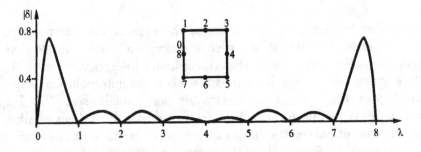

Figure 6.4.

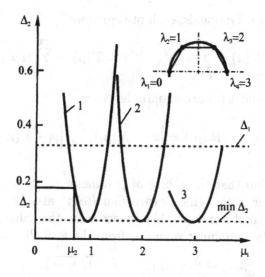

Figure 6.5.

1 shows the variation of the error Δ_2 when the value of μ_2 for node 2 differs from the best value $\lambda_2 = 1$. The values of μ for the remaining nodes are then the best possible: $\mu_i = \lambda_i$. Curves 2 and 3 correspond to the similar situation for nodes 3 and 4. Clearly, when the values of the parameter μ_i deviate from the best value λ_i there is a sharp increase in the error Δ_2.

2. The parametric approximation

The usual approximation problem consists of constructing a polynomial of degree m

$$y = P(x) = \sum_{k=0}^{m} t_k x^k,$$

the graph of which is as close as possible to a given set consisting of n points (x_i, y_i), $i = \overline{1, n}$. In approximation by the method of least squares the measure of closeness is taken as the mean -square deviation

$$\sum_{i=1}^{n} [y_i - P(x_i)]^2.$$

We will consider the problem of parametric approximation using a parameter μ and let the point (x_i, y_i) correspond to the value $\mu = \mu_i$, $i = \overline{1, n}$. Then the problem of parametric approximation consists of constructing two polynomials, each of degree m:

$$x = X(\mu) = \sum_{k=0}^{m} a_k \mu^k, \qquad y = Y(\mu) = \sum_{k=0}^{m} b_k \mu^k, \qquad (6.13)$$

which give the smallest mean-square deviation

$$R = \sum_{i=1}^{n} R_i(\mu_i), \qquad R_i(\mu_i) = \left[x_i - X(\mu_i)\right]^2 + \left[y_i - Y(\mu_i)\right]^2. \qquad (6.14)$$

Bearing in mind that the choice of parameter is not unique, we will seek a parameter μ for which expression (6.14) also has its smallest value. With the first point (x_1, y_1) we associate the value $\mu_1 = 0$, and then compute the remaining values μ_i from the formula

$$\mu_{i+1} = \mu_i + \Delta\mu_i, \qquad i = \overline{1, \, n-1}, \qquad (6.15)$$

taking $\Delta\mu_i$ in the form

$$\Delta\mu_i = \alpha_i \Delta x_i + \beta_i \Delta y_i. \qquad (6.16)$$

Here $\Delta x_i = x_{i+1} - x_i$, $\Delta y_i = y_{i+1} - y_i$; α_i, β_i are constants, a yet undefined, which in the sense of (6.16) can be regarded as components of a vector in the direction of the straight line along which the parameter μ is measured. To ensure that all the directions are equivalent, we will define them by unit vectors, and so we have

$$\alpha_i^2 + \beta_i^2 = 1, \qquad i = \overline{1, \, n-1}. \qquad (6.17)$$

Thus, the problem reduces to finding the extremum of the function (6.14) under the condition that equation (6.17) hold. We will take the Lagrange function L in the form

$$L = R + \sum_{i=1}^{n-1} \gamma_i(-1 + \alpha_i^2 + \beta_i^2), \qquad (6.18)$$

where γ_i are Lagrange multipliers.

The necessary condition for an extremum of a Lagrange function is that its partial derivatives shall equal zero:

$$\begin{aligned}
\frac{\partial L}{\partial a_k} &= -2 \sum_{i=1}^{n} [x_i - X(\mu_i)]\mu_i^k = 0, \\
\frac{\partial L}{\partial b_k} &= -2 \sum_{i=1}^{n} [y_i - Y(\mu_i)]\mu_i^k = 0, \quad k = \overline{0, \, m},
\end{aligned} \qquad (6.19)$$

$$\frac{\partial L}{\partial \alpha_j} = \sum_{i=j+1}^{n} \frac{\partial R_i}{\partial \mu_j} \Delta x_j + 2\gamma_j \alpha_j = 0,$$

$$\frac{\partial L}{\partial \beta_j} = \sum_{i=j+1}^{n} \frac{\partial R_i}{\partial \mu_j} \Delta y_j + 2\gamma_j \beta_j = 0, \quad j = \overline{1, \, n-1}. \tag{6.20}$$

Equations (6.19) can be used to find the coefficients a_k, b_k of polynomials (6.13). If the last terms in equations (6.20) are taken to the right-hand side and the resulting equations are divided by one another, we obtain the relations

$$\frac{\alpha_j}{\beta_j} = \frac{\Delta x_j}{\Delta y_j}, \tag{6.21}$$

and form (6.17) then have

$$\alpha_j = \frac{\Delta x_j}{\sqrt{(\Delta x_j)^2 + (\Delta y_j)^2}}, \qquad \beta_j = \frac{\Delta y_j}{\sqrt{(\Delta x_j)^2 + (\Delta y_j)^2}}. \tag{6.22}$$

Substituting these expressions into (6.16), we obtain the quantities

$$\Delta \mu_j = \sqrt{(\Delta x_j)^2 + (\Delta y_j)^2}, \tag{6.23}$$

which are the same as (6.10) for $\Delta \lambda_i$. Thus, the best parameter must be taken along the broken line joining the given points (x_i, y_i). We will denote it by λ, as before.

Note that when computing α_j, β_j we took a plus sign on the right-hand sides of equations (6.22), corresponding to the minimum of the Lagrange function. It is easy to see that this is so by computing the second differential of the Lagrange function d^2L and neglecting terms with squares and products of Δx, Δy. The corresponding quadratic form will then take the form

$$2\sum_{i=1}^{n-1} \gamma_i \left[(d\alpha_i)^2 + (d\beta_i)^2 \right]. \tag{6.24}$$

Squaring equations (6.20), after taking their last terms to the right-hand side, and adding, using (6.17) we obtain

$$4\gamma_j^2 = \left(\sum_{i=j+1}^{n} \frac{\partial R_i}{\partial \mu_j} \right)^2 \left[(\Delta x_j)^2 + (\Delta y_j)^2 \right],$$

$$\gamma_j = \pm \frac{1}{2} \left| \sum_{i=j+1}^{n} \frac{\partial R_i}{\partial \mu_j} \right| \left[(\Delta x_j)^2 + (\Delta y_j)^2 \right]^{\frac{1}{2}}.$$

We take the positive square root, corresponding to a positive increment $\Delta\mu_j$ and α_j, β_j computed by formulae (6.23) and (6.22). Thus the quadratic form (6.24) will be positive definite, corresponding to a minimum of the Lagrange function (6.18). Note that the values of the parameter μ_i computed using formulae (6.15) will then be positive.

Obviously, the condition for minimum of the Lagrange function is also sufficient. In fact, if the parameter λ is the length of the broken line described above, the direction of each link in the line will be given by a vector with components computed by formulae (6.22), giving a minimum of the Lagrange function (6.18). We thus have the following theorem.

Theorem 6.1. A necessary and sufficient condition for polynomials (6.13) defining a parametric approximation to give the best approximation of the points (x_i, y_i), $i = \overline{1, n}$, is that the length of the broken line joining those points is taken as the parameter μ.

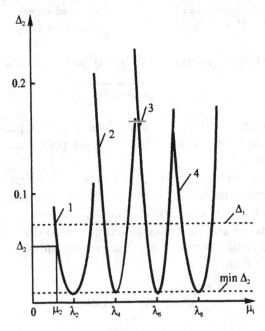

Figure 6.6.

As an example, consider the approximation of points which are situated equal distances apart along the arc of the unit semicircle. The error of the usual approximation Δ_1 and of the parametric approximation Δ_2 will be computed from the formulae

$$\Delta_1 = \left\{ \sum_{i=1}^{n} \left[y_i - P(x_i) \right]^2 \right\}^{\frac{1}{2}},$$

$$\Delta_2 = \left\{ \sum_{i=1}^{n} \left[x_i - X(\lambda_i) \right]^2 \right\}^{\frac{1}{2}} + \left\{ \sum_{i=1}^{n} \left[y_i - Y(\lambda_i) \right]^2 \right\}^{\frac{1}{2}}.$$

Fig. 6.6 shows the results for the approximation of nine points equally spaced over a semicircular arc by parametric polynomials (6.13) of degree six. The error of the usual approximation $\Delta_1 = 0.072$. The error of parametric approximation using the best parameter λ as the length of the broken line joining those points is $\min \Delta_2 = 0.0072$. Curve 1 shows how the error changes as the value of the parameter of the second point μ_2 deviates from its best value λ_2. The values of the parameter μ_i for the remaining points, $(i = 1, 3, 4, \ldots, 9)$, remain equal to the best values λ_i. Curves 2, 3 and 4 show how the error Δ_2 changes in a similar case when the parameter deviates from its best value at points 4, 6 and 8, respectively.

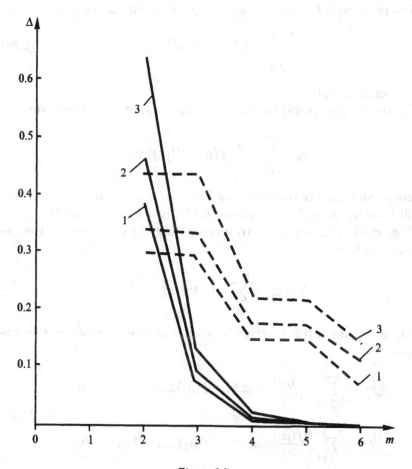

Figure 6.7.

Fig. 6.7 shows how the errors Δ_1 (the dashed line) and Δ_2 (the continuous line) behave as a function of the degree m of the approximating polynomials and the number of points n to be approximated. Curves 1–3 correspond to the values $n = 9$, 15, 33, respectively. On the basis of the recommendations made in [54], we have confined ourselves here to approximating polynomials of up to degree six inclusive. The advantage of parametric approximation is obvious.

3. The continuous approximation

Let the curve to be approximated AB be defined implicitly by means of expression (6.4), and the approximating functions by the parametric relations

$$x = X(\mu), \qquad y = Y(\mu). \tag{6.25}$$

The problem is to find a parameter μ for which the integral

$$\int_{AB} F^2(X(\mu), Y(\mu))\, d\mu \tag{6.26}$$

has its smallest value.

Let the integral (6.26) be the limit as $n \to \infty$ of the integral sum

$$\sigma_n = \sum_{i=1}^{n-1} F^2(X(\mu_i), Y(\mu_i))\Delta\mu_i,$$

where μ_i and $\Delta\mu_i$ are computed by formulae (6.15) and (6.16), and the coefficients α_i, β_i satisfy equations (6.17) (relations (6.5) hold).

The problem then reduces to investigating the extremum of the Lagrange function

$$L = \sigma_n + \sum_{i=1}^{n-1} \gamma_i(-1 + \alpha_i^2 + \beta_i^2).$$

An extremum of this function is possible at points satisfying the conditions

$$\frac{\partial L}{\partial \alpha_j} = \sum_{i=j+1}^{n-1} \frac{\partial f(\mu_i)}{\partial \mu_j}\Delta\mu_i\Delta x_j + f(\mu_j)\Delta x_j + 2\gamma_j\alpha_j = 0,$$

$$\tag{6.27}$$

$$\frac{\partial L}{\partial \beta_j} = \sum_{i=j+1}^{n-1} \frac{\partial f(\mu_i)}{\partial \mu_j}\Delta\mu_i\Delta y_j + f(\mu_j)\Delta y_j + 2\gamma_j\beta_j = 0.$$

Here we have introduced the notation $f(\mu_i) = F^2[X(\mu_i), Y(\mu_i)]$, $i = \overline{1, n-1}$.

If the last terms in equations (6.27) are taken to the right-hand side and the resulting expressions are divided by one another, we obtain relations (6.21). Then, repeating the argument given in the proof of Theorem 6.1 almost word-for-word and taking the limit in the integral sum as $n \to \infty$ we obtain the following theorem.

Theorem 6.2. A necessary and sufficient condition for the integral (6.26) to attain its smallest value is that the parameter μ in the approximating expressions (6.25) is the length λ of the curve (6.4).

As an example, consider the approximation of the parabola defined implicitly in the interval $[0, 1]$ in the form

$$F(x, y) = y - x^2 = 0. \tag{6.28}$$

Representing the function x^2 in the interval $[0, 1]$ in the form of a cosine Fourier series we obtain a solution of the ordinary approximation problem in the form

$$y = \sum_{n=0}^{\infty} a_n \cos(n\pi x), \tag{6.29}$$

$$\text{where} \quad a_0 = \frac{1}{3}, \qquad a_k = \frac{(-1)^k 4}{(\pi k)^2}, \qquad k = 1, 2, , \ldots$$

Clearly, in numerical investigations the summation in (6.29) must be restricted to some finite value of n, equal to m. The error of the usual approximation Δ_1 will be given by a formula of the type: (6.26)

$$\Delta_1 = \int_0^1 \left(\sum_{n=0}^{m} a_n \cos(n\pi x) - x^2 \right)^2 dx. \tag{6.30}$$

We will find the functions (6.25) defining the parametric approximation by integrating the Cauchy problem (6.8) with the function $F(x, y)$ defined in the form (6.28), and with zero initial conditions $x(0) = y(0) = 0$. The error of parametric approximation Δ_2 will be computed by the formula of the type (6.26):

$$\int_{AB} \left[Y(\lambda) - X^2(\lambda) \right]^2 d\lambda. \tag{6.31}$$

The integrals (6.30) and (6.31) were computed numerically by the rectangle method, the integral (6.31) being found when the Cauchy problem

(6.8) was integrated. The value of the integral (6.30) $\Delta_1 = 0.167 \cdot 10^{-3}$, corresponding to $m = 5$, stabilized when the computation step was 0.002. For $m = 10$ we have $\Delta_1 = 0.24 \cdot 10^{-4}$, and for $m = 20$ $\Delta_1 = 0.36 \cdot 10^{-5}$.

As might have been expected, the integral (6.31), computed with the same step 0.002, was equal to $\Delta_2 = 0.341 \cdot 10^{-12}$, which is considerably smaller than the values obtained for Δ_1 above.

Thus, a parametric approximation with the choice of the best parameter possesses a number of advantages over the conventional method of approximation.

Chapter 7

NONLINEAR BOUNDARY VALUE PROBLEMS FOR ORDINARY DIFFERENTIAL EQUATIONS

In previous chapters the examples of the best parametrization were considered for problems, the solutions of which were the simplest one-parameter sets, i.e., curves in Euclidean space. Here we consider a more complicated case of nonlinear boundary value problem for ordinary differential equations (ODE) with a parameter. Solution of this problem is a one-parameter set of curves.

There are various approaches to solution of nonlinear boundary value problems. The most widely used methods are projection and variation methods of the Bubnov and Ritztype and as well difference and variational difference methods such as finite difference and finite elements methods. In all these methods a nonlinear boundary value problems are reduced to systems of nonlinear algebraic or transcendental equations with a parameter and the problems can be solved by a direct application of the parametric continuation algorithm developed in chapter 1. So we do not considered here such algorithms.

Another approach involves local linearization of nonlinear boundary value problem with respect to a parameter. Within the framework of the continuation method this approach is implemented by applying the continuation procedure directly to original equations. The first step in the direction of using the continuation procedure this way was made by Vlasov and Petrov who formulated the incremental loading method [91].

When such approach is used for constructing the solution set of nonlinear boundary value problem with parameter the problem is reduced to a sequence of one-dimension linear boundary value problems which are conveniently solved by shooting type methods. Several versions of the shooting method, which ensure high accuracy and reasonably economy of the solution techniques, have now been worked out. We shall use the discrete orthogonal shooting method of the Godunov [46, 7]. Some modifications of traditional shooting method algorithm will be made to ensure its effectiveness in algorithms of continuation with respect to a parameter. The modifications (and their necessity) will be explained on the example of the initial parameter method which is an important part of the orthogonal shooting method.

1.   The equations of solution continuation for nonlinear one-dimensional boundary value problems

We consider a boundary value problem for a system of nonlinear ODE with a parameter p

$$y' = F(x, y, p), \quad Ay(x_1) = a, \quad By(x_2) = b,$$

$$x \in {}^1, \quad y : {}^1 \to {}^n, \quad F : {}^{n+2} \to {}^n, \quad a \in {}^m, \quad b \in {}^{n-m}. \tag{7.1}$$

Here we introduce the following notation: A is a rectangular nonsingular matrix of size $m \times n$ $(m < n)$; B is a rectangular nonsingular matrix of size $l \times n$, where $l = n - m$; $y' = dy/dx$.

Assume that boundary value problem (7.1) has a solution for some range of the parameter p and also assume that for a certain value $p = p_0$ in this range the solution $y_{(0)}$ is known, i.e.,

$$y|_{p=p_0} = y_{(0)}. \tag{7.2}$$

In line with the basic idea of the continuation we assume the unknown vector function y and the parameter p to be continuous differentiable functions of certain parameter μ , the meaning of which will be determined later

$$y = y(x, \mu), \qquad p = p(\mu). \tag{7.3}$$

Since μ does not enter explicitly into the boundary value problem (7.1) we have right to choose a convenient the reference point for μ. We choose it, therefore, so that $\mu = 0$ corresponds to the known solution (7.2), i.e.,

$$y(x, 0) = y_{(0)}, \qquad p(0) = p_0. \tag{7.4}$$

The derivatives of y and p with respect to the parameter μ are denoted by corresponding capital letters

$$\frac{dy}{d\mu} = Y, \qquad \frac{dp}{d\mu} = P. \tag{7.5}$$

This expressions together with initial conditions (7.4) form the Cauchy problem with respect to the parameter μ. They need to be supplemented by relations defining Y and P. These relations are obtained by differentiating the boundary value problem (7.1) with respect to μ. As a result we have a linear boundary value problem

$$Y' = L(y,p)Y + PM(y,p), \quad AY(x_1) = 0, \quad BY(x_2) = 0. \qquad (7.6)$$

Here $L(y,p) = [L_{ij}(y,p)], \quad M(y,p) = (M_1(y,p),\ldots,M_n(y,p))^T$ are matrix and vector whose components are determined by the relations

$$L_{ij} = \frac{\partial F_i}{\partial y_j}, \quad M_i = \frac{\partial F_i}{\partial p}, \quad i,j = \overline{1,\,n}. \qquad (7.7)$$

Note that the value $P = dp/d\mu$ in the equation (7.6) is also unknown.

One of the methods for solving a linear boundary value problem is the discrete orthogonal shooting method of the Godunov [46]. It is stable and highly efficient. However, using this method within the algorithms of the continuation method with respect to the best parameter requires modifications of the method. To avoid obscuring the essence of these modifications by details connected with discrete orthogonalization of solutions we consider first the solution with the help of initial parameter method which is an essential part of the orthogonal shooting method.

The initial parameter method presumes the solution of the linear boundary value problem (7.6) to be represented in the form

$$Y = C_1 Y^{(1)} + C_2 Y^{(2)} + \ldots + C_l Y^{(l)} + PY^{(l+1)}. \qquad (7.8)$$

Here $l = n - m$; $C_1, C_2, \ldots, C_l$ are arbitrary constants and the vector functions $Y^{(1)}(x), Y^{(2)}(x), \ldots, Y^{(l+1)}(x)$ are linearly independent solutions of the following initial homogeneous problem

$$Y' = LY, \quad AY(x_1) = 0, \quad (Y(x_1) \neq 0). \qquad (7.9)$$

The vector function $Y^{(l+1)}(x)$ is a solution of the nonhomogeneous problem

$$Y' = LY + M, \quad Y(x_1) = 0. \qquad (7.10)$$

The representation (7.8) sets up a correspondence between a functional space of solutions (7.6) satisfying the boundary condition

$AY(x_1) = 0$ and $l + 1$ – dimensional vector space $^{l+1} : \{C_1, \ldots, C_l, P\}$. In other words, for any vector $C = (C_1, \ldots, C_l, P)^T \in {}^{l+1}$ the expression (7.8) will always be a solution of the problem

$$Y' = LY + PM, \qquad AY(x_1) = 0. \tag{7.11}$$

The initial parameter method determines a vector $C \in {}^{l+1}$ such that the function Y (7.8) is a solution of the boundary value problem (7.6), i.e., this function also satisfies the condition $BY(x_2) = 0$. Let us construct a matrix D with its size $n \times (l + 1)$

$$D = [Y^{(1)}(x_2) \quad Y^{(2)}(x_2) \quad \ldots \quad Y^{(l+1)}(x_2)]. \tag{7.12}$$

of vectors $Y^{(i)}(x)$ $(i = \overline{1, \, l + 1})$. These vectors are the values of the vector functions $Y^{(i)}$ for $x = x_2$,

After that the condition $BY(x_2) = 0$ leads to the equation

$$BDC = JC = 0, \qquad J = BD. \tag{7.13}$$

This equation has the same meaning for the nonlinear boundary value problem (7.1) as the condition equation (1.35) for the system of nonlinear equations (1.28). Indeed, because of the representation (7.8) the vector $C \in {}^{l+1}$ determined from equation (7.13) corresponds to vector function $Y(x)$ and parameter P. The latter are the right hand sides of the continuation equations (7.5), i.e., the correspondence

$$\left\{ \frac{dy}{d\mu} = Y, \quad \frac{dp}{d\mu} = P \right\} \to C. \tag{7.14}$$

takes place.

This is why we have denoted the matrix BD by J, to show its connection with the matrix of continuation equations (1.35). Since the matrix B is of the size $l \times n$ and the matrix D of size $(l + 1) \times n$ the size of the matrix J is $l \times (l + 1)$. Thus, the equation (7.13) represents a system of l homogeneous linear algebraic equations with respect to $l + 1$ unknown components of vector $C = (C_1, \ldots, C_l, P)^T \in {}^{l+1}$. By the meaning of continuation process the vector C is a function of parameter μ, i.e., $C = C(\mu)$.

Let us form a vector $c \in {}^{l+1}$ such that

$$\frac{dc}{d\mu} = C. \tag{7.15}$$

For instance, it can easily be constructed as an integral of the form

$$c = \int_0^\mu C(\mu)d\mu. \tag{7.16}$$

Correspondence (7.14) can then be represented as

$$\left\{ \frac{dy}{d\mu} = Y, \quad \frac{dp}{d\mu} = P \right\} \to \left\{ \frac{dc}{d\mu} = C \right\}. \tag{7.17}$$

From this it can be seen that in additional to the correspondence $\{Y,\ P\} \to C$ the representation (7.8) sets up another correspondence

$$\{y,\ p\} \to c.$$

Thus, the functional space of solutions of the nonlinear boundary value problem (7.1) determined by the parameter p is mapped onto the set $c(\mu)$ that, by virtue continuity of $C(\mu)$ and expression (7.16), is a smooth curve K in $^{l+1}$. We determine the parameter μ locally at each point of the curve K in just the same way as in chapter 1. I.e., we assume that

$$d\mu = \alpha \cdot dc, \quad \alpha \in {}^{l+1}, \quad \alpha \cdot \alpha = 1, \tag{7.18}$$

where unit vector α determines direction of the axis along which the parameter μ is measured. Taking into account the expression (7.15), equations (7.13) and (7.18) can be written as

$$J\frac{dc}{d\mu} = 0,$$
$$\alpha \cdot \frac{dc}{d\mu} = 1. \tag{7.19}$$

This system coincides exactly with the system (1.37). So, repeating all reasoning of the chapter 1 developed in the process of proving of the theorem 1, we come to the following conclusion:

the best conditionality of the system (7.19) will be provided if and only if the parameter μ is a parameter of the length λ that measured along the curve K in $^{l+1}$.

This enable us to use all algorithms developed in chapter 1 for the process of continuation.

Let us consider now the incremental process of Lahaye for solving the boundary value problem (7.1). Then, on k – th step with respect to continuation parameter (as $\mu = \mu_k$), it is necessary to formulate the algorithm of Newton – Raphson method for the problem (7.1). We use its generalization known as quasilinearization [8]. Then for $(j+1)$ – th iteration the algorithm reduces to solving of the following linear boundary value problem

$$(y_{(k)}^{(j+1)})' = L(y_{(k)}^{(j)}, p_k^{(j)})(y_{(k)}^{(j+1)} - y_{(k)}^{(j)})+$$

$$+M(y_{(k)}^{(j)}, p_k^{(j)})(p_k^{(j+1)} - p_k^{(j)}) + F(y_{(k)}^{(j)}, p_k^{(j)}), \qquad (7.20)$$

$$Ay_{(k)}^{(j+1)}(x_1) = a, \qquad By_{(k)}^{(j+1)}(x_2) = b, \qquad j = 1, 2, \ldots$$

Such form of the quasi linearization equation emphasizes the fact that if the iterative process converges, i.e., $y_{(k)}^{(j)} \to y_{(k)}$ and $p_k^{(j)} \to p_k$ as $j \to \infty$, it converges to the solution of the original boundary value problem. Indeed, in that case the first two terms on the right hand side of equation (7.20) tend to zero. In this limit the equation (7.20) transforms into the original one (7.1).

It is convenient to rewrite the problem (7.20) in the form

$$(y_{(k)}^{(j+1)})' = L_{(k)}^{(j)} y_{(k)}^{(j+1)} + p_k^{(j)} M_{(k)}^{(j)} + \Phi_{(k)}^{(j)},$$

$$Ay_{(k)}^{(j+1)}(x_1) = a, \qquad By_{(k)}^{(j+1)}(x_2) = b, \qquad j = 1, 2, \ldots \qquad (7.21)$$

Here we have used the notation

$$L_{(k)}^{(j)} = L(y_{(k)}^{(j)}, p_k^{(j)}), \qquad M_{(k)}^{(j)} = M(y_{(k)}^{(j)}, p_k^{(j)}),$$

$$\Phi_{(k)}^{(j)} = F(y_{(k)}^{(j)}, p_k^{(j)}) - L(y_{(k)}^{(j)}, p_k^{(j)}) y_{(k)}^{(j)} - p_k^{(j)} M(y_{(k)}^{(j)}, p_k^{(j)}). \qquad (7.22)$$

This form shows that in incremental Lahaye process, i.e., in the discrete solution continuation, a more complicated (in comparison with the problem (7.6) arising in continuous continuation) linear boundary value problem (7.21) has to be solved.

The problem (7.21) has a more complicated nonhomogeneous in the right hand side of the equation and its boundary conditions are also non-homogeneous. Therefore, its solution by the initial parameter method is represented in the form

$$y_{(k)}^{(j+1)} = C_{1(k)}^{(j+1)} y_{(k)}^{(1)(j+1)} + \ldots + C_{l(k)}^{(j+1)} y_{(k)}^{(l)(j+1)}+$$

$$+p_k^{(j+1)} y_{(k)}^{(l+1)(j+1)} + y_{(k)}^{(l+2)(j+1)}. \qquad (7.23)$$

Here $y_{(k)}^{(i)(j+1)}$ $(i = \overline{1, l})$ are vector-functions, which are linearly independent solutions of the homogeneous problem

$$y' = L_{(k)}^{(j)}y, \qquad Ay(x_1) = 0, \qquad (y(x_1) \neq 0). \tag{7.24}$$

The vector function $y_{(k)}^{(l+1)(j+1)}$ is constructed as a solution of the nonhomogeneous problem

$$y' = L_{(k)}^{(j)}y + M_{(k)}^{(j)}, \qquad y(x_1) = 0. \tag{7.25}$$

Finally, $y_{(k)}^{(l+2)(j+1)}$ represents a solution of the following nonhomogeneous problem

$$y' = L_{(k)}^{(j)}y + \Phi_{(k)}^{(j)}, \qquad Ay(x_1) = a. \tag{7.26}$$

As before, we introduce a matrix $D_{(k)}^{(j+1)}$ made up of column vectors of values of the vector functions $y_{(k)}^{(i)(j+1)}$ $(i = \overline{1, l+1})$ as $x = x_2$

$$D_{(k)}^{(j+1)} = \left[y_{(k)}^{(1)(j+1)}(x_2) \quad y_{(k)}^{(2)(j+1)}(x_2) \quad \dots \quad y_{(k)}^{(l+1)(j+1)}(x_2) \right]. \tag{7.27}$$

We also introduce a vector $C_{(k)}^{(j+1)} = (C_{1(k)}^{(j+1)}, \dots, C_{l(k)}^{(j+1)}, p_k^{(j+1)})^T$ made up of l constants of integration and parameter. Then vector function $y_{(k)}^{(j+1)}(x)$ (7.23) is a solution of the boundary value problem (7.20) if it satisfies the second boundary condition $By_{(k)}^{(j+1)}(x_2) = b$ which, taking into account the introduced notation, reduces to the linear algebraic equation

$$BD_{(k)}^{(j+1)}C_{(k)}^{(j+1)} + By_{(k)}^{(l+2)(j+1)}(x_2) = b. \tag{7.28}$$

This equation can be rewritten as follows

$$J_{(k)}^{(j+1)}C_{(k)}^{(j+1)} = d_{(k)}^{(j+1)}, \tag{7.29}$$

where $J_{(k)}^{(j+1)} = BD_{(k)}^{(j+1)}$, $d_{(k)}^{(j+1)} = b - By_{(k)}^{(l+2)(j+1)}(x_2)$. $\tag{7.30}$

The equation (7.29) represents a system of l nonhomogeneous linear algebraic equations in $l+1$ unknown components of the vector $C_{(k)}^{(j+1)}$. Its solution may be sought in the form

$$C_{(k)}^{(j+1)} = a_{(k)}^{(j+1)} C_{0(k)}^{(j+1)} + g_{(k)}^{(j+1)}. \qquad (7.31)$$

Here $g_{(k)}^{(j+1)}$ is a particular solution of the nonhomogeneous equation (7.29) and $C_{0(k)}^{(j+1)}$ is a general solution of the homogeneous equation

$$J_{(k)}^{(j+1)} C_{0(k)}^{(j+1)} = 0. \qquad (7.32)$$

This equation has a one - parameter subspace of solutions in $^{l+1}$: $\{C_1, \ldots, C_l, p\}$. We will suppose that $C_{0(k)}^{(j+1)}$ is the unit vector of this subspace. Coefficient $a_{(k)}^{(j+1)}$ in representation (7.31) is not determined because the solution of homogeneous equation can be determined only up to an arbitrary multiplier.

As in the previous analysis, we take advantage of the opportunities provided by the interpretation of the solution representation (7.23) as the mapping of the functional vector space $\{y, \ p\}$ onto the $(l+1)$ – dimensional vector space $^{l+1}$: $\{C_1, \ldots, C_l, p\}$. By virtue of this mapping the following linear boundary value problem corresponds to the equation (7.32)

$$(Y_{(k)}^{(j+1)})' = L_{(k)}^{(j)} Y_{(k)}^{(j+1)} + P_k^{(j)} M_{(k)}^{(j)},$$

$$AY_{(k)}^{(j+1)}(x_1) = 0, \qquad BY_{(k)}^{(j+1)}(x_2) = 0. \qquad (7.33)$$

If iterative process converges, i.e.,

$$y_{(k)}^{(j)} \to y_{(k)}, \quad p_k^{(j)} \to p_k, \quad Y_{(k)}^{(j)} \to Y_{(k)}, \quad P_k^{(j)} \to P_k,$$

as $j \to \infty$, the problem (7.33) converges to the following one

$$Y_{(k)}' = L(y_{(k)}, p_k) Y + P_k M(y_{(k)}, p_k),$$

$$AY_{(k)}(x_1) = 0, \qquad BY_{(k)}(x_2) = 0. \qquad (7.34)$$

Apart from notation, this boundary value problem coincides with the boundary value problem (7.6). Thus, the iterative process for vector $C_{0(k)}^{(j)}$ converges to $C_{(k)} = C(\mu_k)$ which is the tangent unit vector to the curve K. The solution set of the boundary value problem (7.1) is mapped onto the K in $^{l+1}$.

The arbitrary coefficient $a_{(k)}^{(j+1)}$ in solution (7.31) can be chosen at each iteration such that the mapping of the iterative process (7.21) onto the space $^{l+1}$ has the same properties as the iterative processes for discrete continuation, considered in section 1.5.

Thus, a condition equivalent to condition (1.85) must require the orthogonality of the correction vectors

$$\delta C_{(k)}^{(j+1)} = C_{(k)}^{(j+1)} - C_{(k)}^{(j)}$$

to the unit vector $C_{0(k-1)}$ of the tangent to the curve $K \in {}^{l+1}$ at the previous point. This condition is written as

$$C_{0(k-1)} \cdot \delta C_{(k)}^{j+1} = 0. \tag{7.35}$$

Substituting in (7.35) the solution (7.31) gives $a_{(k)}^{(j+1)}$ as

$$a_{(k)}^{(j+1)} = -\frac{C_{0(k-1)} \cdot \delta C_{(k)}^{j+1}}{C_{0(k-1)} \delta C_{0(k)}^{j+1}}. \tag{7.36}$$

The geometry of this iterative process in $^{l+1}$ is the same as shown in Fig. 1.10.

Similarly, other conditions formulated in section 1.5 and shown in Fig. 1.11, 1.12. can be used to determine $a_{(k)}^{(j+1)}$.

2. The discrete orthogonal shooting method

As shown in previous section the algorithms of the solution parametric continuation for nonlinear boundary value problems involve the solution of linearized boundary value problems (7.6), (7.21) at each step. But using initial parameter method for solving these problems usually encounters certain difficulties connected with rapidly decreasing and increasing solutions, especially for stiff systems of ODEs. This makes systems (7.13), (7.29) ill conditioned, which is also explained by the fact that matrix J is near singular. One of the most efficient ways of overcoming these difficulties is the use of the discrete orthogonal shooting method developed by Godunov [46]. For realization of continuation algorithm it is necessary to solve linear boundary value problems (7.6), (7.21) containing a parameter to be determined in free terms. It is a distinction from traditional version of orthogonal shooting method presented in [7, 49, 59]. This requires a certain modernization of the well known algorithm for the orthogonal shooting method.

We now consider the linear boundary value problem

$$y' = L(x)y + pM(x) + \Phi(x), \qquad Ay(x_0) = a, \quad By(x_N) = b. \quad (7.37)$$

Here $y(x) = (y_1(x), \ldots, y_n(x))^T$ is unknown n – dimensional vector function; $L(x) = [L_{ij}(x)](i, j = \overline{1, n})$ is a given square matrix function of order n; $M(x) = (M_1(x), \ldots, M_n(x))^T, \Phi(x) = (\Phi_1, \ldots, \Phi_n)^T$ are a given n – dimension vector functions; A is a given rectangular matrix of size $m \times n (n > m, rank A = m)$; B is a given rectangular matrix of the size $l \times n, l = n - m (rank B = l)$; $a = (a_1, \ldots, a_n)^T$, $b = (b_1, \ldots, b_n)^T$ are a given vectors; p is a parameter; x_0, x_N are coordinates of the initial and final points of interval where the solution of the boundary value problem (7.37) is sought.

Before considering the algorithm for the orthogonal shooting method, let us focus our attention on the property of the general solution of equation (7.37) satisfying the condition $Ay(x_0) = 0$. This solution is represented as

$$y(x) = C_1 y^{(1)} + C_2 y^{(2)} + \ldots + C_l y^{(l)} + p y^{(l+1)} + y^{(l+2)}, \quad l = m - n. \quad (7.38)$$

Here $y^{(j)} = (y_1^{(j)}, \ldots, y_n^{(j)})^T (j = \overline{1, l})$ are linearly independent vector functions which are solutions of the homogeneous problem

$$y' = L(x)y, \qquad Ay(x_0) = a \ (y(x_0) \neq 0); \qquad (7.39)$$

$y^{(l+1)} = (y_1^{(l+1)}, \ldots, y_n^{(l+1)})^T$ is a vector function which is a particular solution of the problem

$$y' = L(x)y + pM(x), \qquad Ay(x_0) = a; \qquad (7.40)$$

$y^{(l+2)} = (y_1^{(l+2)}, \ldots, y_n^{(l+2)})^T$ is a vector function such that it is a particular solution of the problem

$$y' = L(x)y + \Phi(x), \qquad Ay(x_0) = a; \qquad (7.41)$$

$C_1, \ldots, C_l$ are arbitrary constants; p is a parameter.

We introduce the vector $C = (C_1, \ldots, C_l, p, 1)^T \in {}^{l+2}$ and the matrix $U(x)$ such that its columns are vector functions $y^{(j)} (j = \overline{1, l+2})$,

$$U(x) = [U_{ij}(x)] = [y^{(1)}(x) \quad y^{(2)}(x) \quad \cdots \quad y^{(l+2)}(x)], \qquad (7.42)$$

where $U_{ij}(x) = y_i^{(j)}(x)$, $i = \overline{1, n}$, $j = \overline{1, l + 2}$.

This matrix will be called the general solution matrix. Then the solution of the problem (7.38) is represented in matrix form

$$y(x) = UC. \tag{7.43}$$

We introduce a nonsingular square matrix Q of the order $l + 2$ with the following structure

$$Q = \begin{bmatrix} q_{11} & \cdots & q_{1l} & q_{1l+1} & q_{1l+2} \\ \vdots & \ddots & \vdots & \vdots & \vdots \\ q_{l1} & \cdots & q_{ll} & q_{ll+1} & q_{ll+2} \\ 0 & \cdots & 0 & 1 & 0 \\ 0 & \cdots & 0 & 0 & 1 \end{bmatrix}, \tag{7.44}$$

It is easy to see that the matrix U^* obtained as

$$U^* = UQ, \tag{7.45}$$

has the same structure as the matrix U. Its first l columns are linearly independent combinations of the columns of U matrix, i.e., they are linearly independent solutions of the homogeneous problem (7.39). The last two columns of U^* are particular solutions of the problems (7.40) and (7.41). Hence, the vector function

$$y^*(x) = U^*C^*, \tag{7.46}$$

where $C^* = (C_1^*, \ldots, C_l^*, p, 1)^T$ is also a general solution of the equation (7.37) satisfying the condition $Ay(x_0) = a$ with arbitrary values of the constants $C_1^*, \ldots, C_l^*$. Thus, U^* is also a general solution matrix. If $y(x) \equiv y^*(x)$ the vectors C and C^* are related, in view (7.45), by the equation

$$QC = C^*. \tag{7.47}$$

We consider now the algorithm for orthogonal shooting method. To do this, we divide the interval of integration $x_0 \le x \le x_N$ into N segments. Let us denote the coordinates of segments boundaries as $x_0, x_1, \ldots, x_N$ and let us number the N segments from left to right. In the first segment $x_0 \le x \le x_1$ we construct the general solution (7.38) of the system (7.37) satisfying the condition $Ay(x_0) = a$. We first obtain the solution of homogeneous problem (7.39) and denote it as $y_{(1)}^{(j)} (j = \overline{1, l})$. Here and below the subscript (i) in parentheses indicates a function or a quantity belonging to the i – th segment. To do this, we take l orthonormal solutions $\xi_{(1)}^{(j)} \in \,^n (j = \overline{1,\,l})$ of the equation $A\xi = 0$ and using them as a initial solutions we construct l solutions of the initial value problem $y' = L(x)y$ with any numerical method (Runge - Kutta, Adams - Stormer, etc.), i.e., we obtain $y_{(1)}^{(j)}(j = \overline{1,\,l})$ as a solution of the initial value problems

$$(y_{(1)}^{(j)})' = L(x)y_{(1)}^{(j)}, \quad y_{(1)}^{(j)}(x_0) = \xi_{(1)}^{(j)}, \quad j = \overline{1,\,l}, \quad x_0 \le x \le x_1. \quad (7.48)$$

The solutions $y_{(1)}^{(l+1)}$ and $y_{(1)}^{(l+2)}$ we obtain as a solutions of the following initial value problem

$$(y_{(1)}^{(l+1)})' = L(x)y_{(1)}^{(l+1)} + M(x), \quad y_{(1)}^{(l+1)}(x_0) = \xi_{(1)}^{(l+1)} = 0, \quad (7.49)$$

$$(y_{(1)}^{(l+2)})' = L(x)y_{(1)}^{(l+2)} + \Phi(x), \quad y_{(1)}^{(l+2)}(x_0) = \xi_{(1)}^{(l+2)}. \quad (7.50)$$

Here $\xi_{(1)}^{(l+2)} \in \,^n$ is a vector that is a particular solution of the system $A\xi = a$ and it is orthogonal to the vectors $\xi_{(1)}^{(j)}(j = \overline{1,\,l})$ but not normalized.

We will consider how the vectors $\xi_{(1)}^{(j)}(j = \overline{1,\,l+2})$ can be constructed later.

Now the general solution of the problem (7.37) satisfying the condition $Ay(x_0) = a$ on the first segment can now be represented in the form

$$y_{(1)}(x) = U_{(1)}(x)C_{(1)}. \quad (7.51)$$

Here $U_{(1)}(x)$ is the matrix function (7.42) and $C_{(1)} = (C_{(1)1}, \ldots, C_{(1)l}, p, 1)^T$ is a vector of arbitrary constants on the first segment.

At the left end of the first segment (as $x = x_0$) the matrix function $U_{(1)}(x)$ assumes the value

$$U_{(1)}(x_0) = \xi_{(1)}, \quad \xi_{(1)} = (\xi_{(1)}^{(1)}, \ldots, \xi_{(1)}^{(l+2)})^T. \quad (7.52)$$

Here $\xi_{(1)}$ is a matrix whose columns are the orthogonal vectors $\xi_{(1)}^{(j)}(j = \overline{1,\,l+2})$ introduced above.

At the right end of the first segment (as $x = x_1$) the value of the matrix $U_{(1)}(x)$ is denoted as

$$U_{(1)}(x_1) = \Psi_{(1)} = (\Psi_{(1)}^{(1)}, \ldots, \Psi_{(1)}^{(l+2)})^T,$$
$$\Psi_{(1)}^{(j)} = y_{(1)}^{(j)}(x_1), \quad j = \overline{1, \, l+2}. \tag{7.53}$$

The columns of the matrix $U_{(1)}(x_0) = \xi_{(1)}$ form an orthogonal system of vectors in ^{n} but the columns of the matrix $U_{(1)}(x)$ deviate from orthogonal system as x increases. This deviation grows especially rapidly if equation (7.37) has rapidly increasing and decreasing homogeneous solutions. In this case the vector $y_{(1)}^{(j)}(x)(j = \overline{1, \, l+2})$ may become almost linearly dependent for large x. The idea of discrete orthogonal shooting method is to halt the integration process and pass to another general solution in the next segment whilst these deviations are not too large. This solution must be such that its vectors of homogeneous and particular solutions are orthogonal as $x = x_1$, i.e., at the beginning of the next segment. To accomplish this we orthogonalize the columns of the matrix (7.53) by the Gramm – Schmidt process and in addition normalize the first l columns that represent the solutions of the homogeneous problem (7.39) in Ψ. The vectors $\Psi_{(1)}^{(l+1)}$ and $\Psi_{(1)}^{(l+2)}$ are only orthogonalized to $\Psi_{(1)}^{(j)}(j = \overline{1, \, l})$ without normalization. Denote the resulting system of vectors by $\xi_{(2)}^{(j)}(x)(j = \overline{1, \, l+2})$ and form a matrix $\xi_{(2)}$ of them in exactly the same way as the matrix $\xi_{(1)}$ (7.52) is formed of the vectors $\xi_{(1)}^{(j)}(x)(j = \overline{1, \, l+2})$. The result of this operation can be represented as

$$\Psi_{(1)} = \xi_{(2)}\Omega_{(1)}. \tag{7.54}$$

Since the matrix $\Psi_{(1)}$ is orthogonalized in columns the matrix $\Omega_{(1)}$ is upper triangular matrix. Taking into account the specific features of the process connected with the orthogonalization of the vectors $\Psi_{(1)}^{(l+1)}$ and $\Psi_{(1)}^{(l+2)}$, it is written in the form

$$\Omega_{(1)} = \begin{bmatrix} \omega_{11}^{(1)} & \omega_{12}^{(1)} & \cdots & \omega_{1l}^{(1)} & \omega_{1l+1}^{(1)} & \omega_{1l+2}^{(1)} \\ 0 & \omega_{22}^{(1)} & \cdots & \omega_{2l}^{(1)} & \omega_{2l+1}^{(1)} & \omega_{2l+2}^{(1)} \\ \vdots & \vdots & \ddots & \vdots & \vdots & \vdots \\ 0 & 0 & \cdots & \omega_{ll}^{(1)} & \omega_{ll+1}^{(1)} & \omega_{ll+2}^{(1)} \\ 0 & 0 & \cdots & 0 & 1 & 0 \\ 0 & 0 & \cdots & 0 & 0 & 1 \end{bmatrix}, \tag{7.55}$$

The vectors $\xi_{(2)}^{(j)}(j = \overline{1, l})$ are expressed in terms of the vectors $\Psi_{(1)}^{(j)}(j = \overline{1, l})$ as follows

$$\xi_{(2)}^{(j)} = \frac{1}{\omega_{ij}^{(1)}}\left(\Psi_{(1)}^{(j)} - \sum_{i=1}^{j-1}\omega_{ij}^{(1)}\xi_{(2)}^{(j)}\right), \quad j = \overline{1, l}, \tag{7.56}$$

where $\omega_{ij}^{(1)} = \Psi_{(1)}^{(j)}\xi_{(2)}^{(i)}, \quad i < j,$

$$\omega_{jj}^{(1)} = \left[\Psi_{(1)}^{(j)}\Psi_{(1)}^{(j)} + \sum_{i=1}^{j-1}(\omega_{jj}^{(1)})^2\right]^{1/2}. \tag{7.57}$$

The vectors $\xi_{(2)}^{(l+1)}$ and $\xi_{(2)}^{(l+2)}$ are not normalized and they are calculated by the formula

$$\xi_{(2)}^{(k)} = \Psi_{(1)}^{(k)} - \sum_{i=1}^{l}\omega_{ki}^{(1)}\xi_{(2)}^{(i)}, \quad k = l+1, \quad k = l+2. \tag{7.58}$$

The resulting orthogonal system of vectors $\xi_{(2)}^{(j)}(j = \overline{1, l+2})$ is taken as a starting system for constructing the solution $y_{(2)}^{(j)}(j = \overline{1, l+2})$ making up the general solution of equation (7.37) in the second segment. I.e., just as for the first segment we construct solutions of the following initial value problems

$$(y_{(2)}^{(j)})' = L(x)y_{(2)}^{(j)}, \quad y_{(2)}^{(j)}(x_1) = \xi_{(2)}^{(j)}, \quad j = \overline{1, l}, \quad x_1 \le x \le x_2. \tag{7.59}$$

$$(y_{(2)}^{(l+1)})' = L(x)y_{(2)}^{(l+1)} + M(x), \quad y_{(2)}^{(l+1)}(x_1) = \xi_{(2)}^{(l+1)} = 0. \tag{7.60}$$

$$(y_{(2)}^{(l+2)})' = L(x)y_{(2)}^{(l+2)} + \Phi(x), \quad y_{(2)}^{(l+2)}(x_1) = \xi_{(2)}^{(l+2)}. \tag{7.61}$$

Then, the general solution is formed as

$$y_{(2)}(x) = U_{(2)}(x)C_{(2)}. \tag{7.62}$$

$$U_{(2)}(x) = [y_{(2)}^{(1)} \ \cdots \ y_{(2)}^{(l+2)}], \qquad C_2 = (C_{(2)}^{(1)}, \dots, C_{(2)}^{(l)}, p, 1)^T. \qquad (7.63)$$

By (7.59) – (7.61), we have

$$U_{(2)}(x_1) = \xi_{(2)} = (\xi_{(2)}^{(1)}, \dots, \xi_{(2)}^{(l+2)})^T. \qquad (7.64)$$

From relations (7.52), (7.54), (7.62), (7.64) it follows that

$$U_{(1)}(x) = U_{(2)}(x)\Omega_{(1)}. \qquad (7.65)$$

By the property considered above, $y_{(2)}(x)$ (7.62) is the general solution of equation (7.37) satisfying condition $Ay_{(2)}(x_0) = a$. Since $y_{(1)}(x)$ (7.51) is the general solution of the same problem, the vectors of arbitrary constants $C_{(1)}$ and $C_{(2)}$ are related by the equation

$$C_{(2)} = \Omega_{(1)} C_{(1)}. \qquad (7.66)$$

Proceeding in exactly the same way to segments $3, 4, \dots, N$, we construct in the segments the general solutions $y_{(i)}(x)$ of the equation (7.37) satisfying the condition $Ay_{(i)}(x_0) = a$ as

$$y_{(i)}(x) = U_{(i)}(x)C_{(i)}, \qquad i = \overline{1, \ N}. \qquad (7.67)$$

Since all $y_{(i)}(x)(i = \overline{1, \ N})$ are the general solutions of the same problem the vectors $C_{(i)}$ and $C_{(i+1)}$ are related by the equation

$$C_{(i+1)} = \Omega_{(i)} C_{(i)}, \qquad i = \overline{1, \ N-1}, \qquad (7.68)$$

where $\Omega_{(i)}$ is an orthogonalizing matrix of the form (7.55). It is also such that

$$\Psi_{(i)} = \Omega_{(i)} \xi_{(i+1)}, \quad \Psi_{(i)} = U_{(i)}(x_i),$$
$$\xi_{(i+1)} = U_{(i+1)}(x_i), \quad i = \overline{1, \ N-1}, \qquad (7.69)$$

The vector of constants $C_{(N)}$ is found from the condition at the right end (as $x = x_N$). This condition takes the form $By_{(N)}(x_N) = b$. Taking into account expressions (7.67) it reduces to the following

$$B\Psi_{(N)}C_{(N)} = b, \qquad \Psi_{(N)} = U_{(N)}(x_N). \qquad (7.70)$$

Since the unknowns in the vector $C_{(N)} = (C_{(N)}^{(1)}, \ldots, C_{(N)}^{(l)}, p, 1)^T$ are only first $(l+1)$ components, the equation (7.69) can be represented in the form

$$JC_{(N)} = d. \tag{7.71}$$

Here

$$J = BD, \qquad d = b - B\Psi_{(N)}^{(l+2)},$$

$$D = [\Psi_{(N)}^{(1)} \cdots \Psi_{(N)}^{(l+1)}], \quad C_{(N)} = (C_{(N)}^{(1)}, \ldots, C_{(N)}^{(l)}, p)^T. \tag{7.72}$$

The matrix B is of the size $l \times n$ and the matrix D is of the size $n \times (l+1)$ so that the size of matrix $J = BD$ is equal to $l \times (l+1)$.

The equation (7.71) plays the same role as equation (7.29) in solving the problem (7.37) by the initial parameter method. Therefore, its solution can be sought in the same form as in the section 7.1: $C_{(N)} = aC_{0(N)} + g_{(N)}$, where the coefficient a is determined from additional condition such as the condition (1.85) as example.

After the vector $C_{(N)}$ is found by one method or another, the backward substitution is performed in which the vectors $C_{(i)}$ are determined as the solution of equations (7.66)

$$\Omega_{(i)}C_{(i)} = C_{(i+1)}, \qquad i = N-1, \ldots, 1. \tag{7.73}$$

The solution of the boundary value problem (7.37) is constructed on the segments from the resulting $C_{(i)}$ and from the matrices $U_{(i)}$ obtained in the forward substitution in accordance with (7.67)

$$y_i(x) = U_{(i)}(x)C_{(i)}, \quad x_{i-1} \le x \le x_i, \quad i = N, \ldots, 1. \tag{7.74}$$

We now turn to the question of how the vectors $\zeta^{(i)}(i = \overline{1, l+2})$ can be constructed. Let us discuss this question on the assumption that the principal (left) minor of the matrix A is non-zero. In other cases the modifications to the solution are obvious.

We first construct l linearly independent solutions $\xi^{(i)}(i = \overline{1, l})$ of the homogeneous equation $A\xi = 0$. This equation is represented as

$$a_{11}\xi_1 + \ldots + a_{1m}\xi_m = a_{1m+1}\xi_{m+1} + \ldots + a_{1n}\xi_n,$$

$$\ldots\ldots\ldots\ldots\ldots\ldots\ldots\ldots\ldots\ldots\ldots\ldots\ldots\ldots\ldots\ldots\ldots\ldots\ldots\ldots\ldots \tag{7.75}$$

$$a_{m1}\xi_1 + \ldots + a_{mm}\xi_m = a_{mm+1}\xi_{m+1} + \ldots + a_{mn}\xi_n.$$

By prescribing now l linearly independent combinations of values $\xi^{(i)} (i = \overline{m+1,n})$ entering into the right hand sides of these equations, say, in the form $(1,0,\ldots,0)$, $(0,1,\ldots,0)$, $\ldots$, $(0,0,\ldots,1)$ and solving the system (7.75), we obtain l linearly independent vectors $\xi^{(i)} (i = \overline{1,l})$. On orthogonalizing these vectors by the Gram – Schmidt process we construct l orthonormal solutions $\zeta_{(1)}^{(i)} (i = \overline{1,l})$ of the equation $A\zeta = 0$.

To obtain the vector $\zeta_{(1)}^{(l+2)}$, we first construct an arbitrary particular solution $\xi^{(l+2)} \in {}^n$ of the equation $A\xi = 0$. For this purpose it is sufficient to set the last l of its components equal to zero, i.e., $\xi_{m+1}^{(l+2)} = \ldots = \xi_n^{(l+2)} = 0$ and to find the first m components as a solution of the equations

$$a_{11}\xi_1^{(l+2)} + \ldots + a_{1m}\xi_m^{(l+2)} = a_1,$$

$$\ldots\ldots\ldots\ldots\ldots\ldots\ldots\ldots\ldots\ldots\ldots\ldots \tag{7.76}$$

$$a_{m1}\xi_1^{(l+2)} + \ldots + a_{mm}\xi_m^{(l+2)} = a_m.$$

Orthogonalizing (without normalization) the obtained vector $\xi^{(l+2)} = (\xi_1^{(l+2)}, \ldots, \xi_m^{(l+2)}, 0, \ldots, 0)^T$ to the vectors $\zeta_{(1)}^{(i)} (i = \overline{1,l})$, we construct the required vector $\zeta_{(1)}^{(l+2)}$ in the form

$$\zeta_{(1)}^{(l+2)} = \xi^{(l+2)} - \sum_{i=1}^{l} (\zeta_{(1)}^{(i)} \xi^{(l+2)}) \zeta_{(1)}^{(i)}. \tag{7.77}$$

The vector $\zeta_{(1)}^{(l+1)} = 0$, as assumed in (7.49).

3.   The algorithms for continuous and discrete continuation of the solution with respect to a parameter for nonlinear one - dimensional boundary value problems

To summarize the results of this chapter, we present algorithms for continuous and discrete continuation of the solution with respect to a parameter for nonlinear boundary value problem (7.1). Among the algorithms for continuous continuation we consider Euler method and modified Euler method. A reader can easily construct the algorithms for Runge – Kutta methods and other methods implementing the explicit schemes for integrating the Cauchy problem with respect to a parameter by analogy.

As an example of algorithms for discrete continuation we consider the one connected with additional condition of the form (1.85) or (7.35). Modifications needed for use of other conditions from section 1.5 are obvious.

1). The Euler method.

Consider an algorithms for Euler method of integrating the Cauchy problem with respect to the best parameter λ for the continuation equation

$$\frac{dy}{d\lambda} = Y, \quad \frac{dp}{d\lambda} = P, \quad y|_{\lambda=\lambda_0} = y_{(0)}, \quad p|_{\lambda=\lambda_0} = p_0, \qquad (7.78)$$

where $y(x, \lambda)$ and $p(\lambda)$ form the solution set of nonlinear boundary value problem (7.1).

Assume that the motion along a parameter is performed with a step $\Delta\lambda$, i.e.

$$\lambda_{k+1} = \lambda_k + \Delta\lambda. \qquad (7.79)$$

The superscript or subscript "k" will indicate that a given function or quantity is taken for $\lambda = \lambda_k$.

For determining the right hand sides of the equations (7.78) at each step with respect of the parameter λ it is necessary to solve the linearized boundary value problem (7.6). It will be solved using the orthogonal shooting method. Similar to the section 7.2, the interval $x_0 \leq x \leq x_N$, where the solution of nonlinear boundary value problem is constructed, is assumed to be divided into N segments. The subscript "i" will indicate that a function or a quantity with this subscript corresponds to the i − th segment $x_{i-1} \leq x \leq x_i$.

1. Before starting the process of integration of the Cauchy problem with respect to the parameter λ, it is necessary to specify the initial state defined by initial condition (7.78)

$$k = 0, \quad \lambda_0 = 0, \quad y_{(0)} = y(x, 0),$$
$$p_0 = p(0), \quad \alpha_{(0)} = (0, \dots, 0, 1) \in {}^{l+1}. \qquad (7.80)$$

The row vector $\alpha_{(0)}$ is needed for forming the matrix of the system (7.19) when $\mu = \lambda$. Such a choice of the vector $\alpha_{(0)}$ means that the first step is made with respect to the parameter p.

2. In the orthogonal shooting method it is necessary to construct l orthonormal homogeneous solutions $\zeta^{(j)}(j = \overline{1, l})$ of the algebraic system

of equations

$$A\zeta = 0. \tag{7.81}$$

As pointed out in section 7.2, it is necessary to construct l linearly independent solutions $\xi^{(j)}(j = \overline{1,l})$ of this system first, and then to orthogonalize them by the Gram – Schmidt process. The result can be represented by means of the orthogonalization matrix

$$[\xi^{(1)} \ldots \xi^{(l)}] = [\zeta_{(1)}^{(1)} \ldots \zeta_{(1)}^{(l)}]\Omega_{(0)}.$$

3. Then forward substitution is performed, i.e. the construction of general solutions of the problem

$$\frac{dY_{(k)}}{dx} = L(y_{(k)}, p_k)Y_{(k)} + M(y_{(k)}, p_k), \qquad AY_{(k)}(x_0) = 0. \tag{7.82}$$

by segment $(x_{i-1} \leq x \leq x_i)$.   Orthogonalization matrices $\Omega_{(i)}(i = \overline{1, N-1})$, which relate general solutions for adjacent segments are also constructed.

The linearized boundary value problem of continuous continuation (7.6) is simpler than the problem (7.21) since its boundary conditions are homogeneous ($a = b = 0$) and function $\Phi = 0$. Therefore, the general solution of the problem (7.82) consist of l linearly independent solutions $Y_{(k)}^{(j)}(j = \overline{1,l})$ of the homogeneous problem (7.39) and particular solution $Y_{(k)}^{(l+1)}$ of the nonhomogeneous problem (7.40). The general solution matrices $U_{(i)}^{(k)}(x)(i = \overline{1,N})$ by segments are therefore composed of only $l + 1$ columns

$$U_{(i)}^{(k)}(x) = [Y_{(k)}^{(1)}(x) \ldots Y_{(k)}^{(l+1)}(x)], \qquad i = \overline{1, N}. \tag{7.83}$$

They are constructed successively as solutions of initial value problems of the form (7.48) (as $a = 0$) and (7.49), and are interrelated by

$$U_{(i)}^{(k)}(x_i) = U_{(i+1)}^{(k)}(x_i)\Omega_{(i)}^{(k)}, \qquad i = \overline{1, N-1}. \tag{7.84}$$

Because of the fact that $Y_{(k)}^{(l+2)} \equiv 0$, the matrix $\Omega_{(i)}^{(k)}$ differs from the orthogonalization matrix (7.55) by the absence of the last row and the last column.

As a result of the forward substitution we obtain the general solution matrices by segments

$$U_{(i)}^{(k)}(x), \qquad i = \overline{1, N}, \tag{7.85}$$

the orthogonalization matrices

$$\Omega_{(i)}^{(k)}(x), \qquad i = \overline{1, N-1} \tag{7.86}$$

and the matrix

$$D^{(k)} = U_{(N)}^{(k)}(x_N). \tag{7.87}$$

4. From the condition $BY(x_N) = 0$ we find the vector of arbitrary constants $C_{(N)}^{(k)} = (C_{(N)1}^{(k)}, \ldots, C_{(N)l}^{(k)}, P_k)^T$ which determines the solution of the linearized boundary value problem of continuous continuation. To do that we construct and solve the following system of equations of type (7.19)

$$\begin{bmatrix} J_{(k)} \\ \alpha_{(k-1)} \end{bmatrix} C^{(k)} = \begin{bmatrix} 0 \\ 1 \end{bmatrix}, \tag{7.88}$$

$$\text{where} \quad J_{(k)} = BD^{(k)}. \tag{7.89}$$

using matrix $D^{(k)}$ and vector $\alpha_{(k-1)}$

The obtained vector $C^{(k)}$ should be normalized

$$C_{(N)}^{(k)} = \frac{C^{(k)}}{(C^{(k)} \cdot C^{(k)})^{1/2}}. \tag{7.90}$$

The value P_k is determined as the last component of the vector $C^{(k)}$

$$P_k = C_{(N)l+1}^{(k)}.$$

5. The backward substitution is reduced to constructing the solutions of linearized boundary value problems by segments in the form

$$Y_{(k)}(x) = U_{(i)}^{(k)} C_{(i)}^{(k)}, \qquad x_{i-1} \le x \le x_i, \quad i = N, N-1, \ldots, 1. \tag{7.91}$$

Here $C_{(i)}^{(k)}$ are found as solutions of the systems of equations

$$C_{(i)}^{(k)} \Omega_{(i)}^{(k)} = C_{(i+1)}^{(k)}, \qquad i = N-1, N-2, \ldots, 1. \tag{7.92}$$

6. It is now possible to take a step of Euler method

$$\lambda_{k+1} = \lambda_k + \Delta\lambda, \quad y_{(k+1)} = y_{(k)} + \Delta\lambda Y_{(k)}, \quad p_{k+1} = p_k + \Delta\lambda P_k. \tag{7.93}$$

7. The resulting values of $y_{(k+1)}$ and p_{k+1} are used for next step. The next step is made by repeating the computations beginning with the part 3 for $k+1$, substituting $k+1$ for k, and taking $\alpha_{(k+1)} = C_{(N)}^{(k)}$.

It is well known that Euler method accumulates error, which at each step is of order $O(\Delta\lambda^2)$.

2). Modified Euler method.

The modified Euler method gives an error of order $O(\Delta\lambda^3)$ at each λ - step. We shall consider a version of this method which has already been used in section 1.4, namely (1.4.4). Since in this method all computations at the first half - step completely repeat Euler method, we present them as a flowchart, in which the notation $\{F(x) = 0\}^{-1} \to x$ means that x is obtained as a result of solution of the equation $F(x) = 0$.

1. Specification of the initial state.

$$k = 0, \qquad \lambda_0 = 0, \qquad y_{(0)} = y(x, 0),$$

$$p_0 = p(0), \qquad \alpha_{(0)} = (0, \ldots, 0, 1)^T. \tag{7.94}$$

2. Construction of l orthonormal solutions $\zeta^{(j)} (j = \overline{1, l})$ of the equation $A\zeta = 0$.

3. Determination of the derivatives Y and P at the initial point of the interval $\Delta\lambda$.

3.1. Forward substitution

$$\left\{ \frac{dY_{(k)}}{dx} = L(y_{(k)}, p_k)Y_{(k)} + M(y_{(k)}, p_k), AY_{(k)}(x_0) = 0 \right\}^{-1} \to$$

$$\to U_{(i)}^{(k)}(x)(i = \overline{1, N}), \quad \Omega_{(i)}^{(k)}(i = \overline{1, N-1}). \tag{7.95}$$

3.2. Determination of the vector $C_{(N)}^{(k)}$

$$D^{(k)} = U_{(N)}^{(k)}(x_N), \quad J_{(k)} = BD^{(k)},$$

$$C^{(k)} = \begin{bmatrix} J_{(k)} \\ \alpha_{(k)} \end{bmatrix}^{-1} \begin{bmatrix} 0 \\ 1 \end{bmatrix}, \quad C_{(N)}^{(k)} = \frac{C^{(k)}}{(C^{(k)} \cdot C^{(k)})^{\frac{1}{2}}}, \quad P_k = C_{(N)l+1}^{(k)}. \tag{7.96}$$

3.3. Backward substitution, consecutive solution of the systems

$$\{C_{(i)}^{(k)}\Omega_{(i)}^{(k)} = C_{(i+1)}^{(k)}\}^{-1} \to C_{(i)}^{(k)}, \qquad i = \overline{N-1, 1}; \tag{7.97}$$

construction of solution of the linearized boundary value problem

$$Y_{(k)} = U_{(i)}^{(k)} C_{(i)}^{(k)}, \qquad x_{i-1} \le x \le x_i, \qquad i = \overline{N, 1}. \tag{7.98}$$

4. Construction of the first approximation for the function $y(x)$ and parameter p_{k+1} at the final point of the interval $\Delta\lambda$ at the k - th step

$$y_{(k+1)}^* = y_{(k)} + \Delta\lambda Y_{(k)}, \quad p_{k+1}^* = p_k + \Delta\lambda P_k, \quad \alpha(k) = C_{(N)}^{(k)}. \tag{7.99}$$

5. Computation of the approximate values of the derivatives at the final point of the interval $\Delta\lambda$ at the k - th step.

5.1. Forward substitution

$$\begin{cases} \dfrac{dY_{(k+1)}^*}{dx} = L(y_{(k+1)}^*, p_{k+1}^*)Y_{(k+1)}^* + M(y_{(k+1)}^*, p_{k+1}^*), \\[2mm] AY_{(k+1)}^*(x_0) = 0 \end{cases}^{-1} \to U_{(i)}^{*(k+1)}(x), \tag{7.100}$$

$$i = \overline{1, N}, \quad \Omega_{(i)}^{*(k+1)}, \quad i = \overline{1, N-1}.$$

5.2. Determination of the vector $C_{(N)}^{*(k+1)}$

$$D^{*(k+1)} = U_{(N)}^{*(k+1)}(x_N), \quad J_{(k+1)}^* = BD^{*(k+1)}, \quad P_{k+1}^* = C_{(N)l+1}^{*(k+1)}$$

$$C^{*(k+1)} = \begin{bmatrix} J_{(k)}^* \\ \alpha_{(k)} \end{bmatrix}^{-1} \begin{bmatrix} 0 \\ 1 \end{bmatrix}, \quad C_{(N)}^{*(k+1)} = \frac{C^{*(k+1)}}{(C^{*(k+1)} \cdot C^{*(k+1)})^{\frac{1}{2}}}. \tag{7.101}$$

5.3. Backward substitution

$$\{C_{(i)}^{*(k+1)} \Omega_{(i)}^{*(k+1)} = C_{(i+1)}^{*(k+1)}\}^{-1} \to C_{(i)}^{*(k+1)}, \qquad i = \overline{N-1, 1}; \tag{7.102}$$

$$Y_{(k+1)}^* = U_{(i)}^{*(k+1)} C_{(i)}^{*(k+1)}, \qquad x_{i-1} \le x \le x_i, \quad i = \overline{N, 1}. \tag{7.103}$$

6. A step of the modified Euler method. Computation of the improved value of $y(x)$ and p at the final point of the interval $\Delta\lambda$ at the k - th step

$$\lambda_{(k+1)} = \lambda_{(k)} + \Delta\lambda, \qquad y_{(k+1)} = y_{(k)} + \Delta\lambda(Y_{(k)} + Y_{(k+1)}^*)/2,$$

$$p_{k+1} = p_k + \Delta\lambda(P_{(k)} + P_{(k+1)}^*)/2. \tag{7.104}$$

7. Passing to the next step with respect to λ: $\alpha_{(k+1)} = C_{(N)}^{*(k+1)}$, repetition of the computations beginning with the part 3 replacing k by $k + 1$.

Although the modified Euler method leads to the accumulation of a smaller error that Euler method, the error may still be significant if the number of λ - steps is sufficiently large.

Further reduction of the error can be achieved in two ways. One of these is to increase the accuracy of explicit schemes that can be done by using such methods as Runge – Kutta or Adams – Stormer. Based on these methods the algorithms for continuation of the solution of a nonlinear boundary value problem with respect to a parameter will be similar to those constructed previously. This, however, requires extra computational time and computer storage space. An alternative is to use implicit schemes, i.e., to pass to discrete continuation of solution.

3). Algorithm for discrete continuation of solution.

Here we shall only consider the algorithm which uses additional condition of the form (1.85). The other additional conditions of section 1.5 can be implemented in a similar fashion.

Let us consider the iterative process for $\lambda = \lambda_k$. Below the correspondent index k will be omitted. The index j will indicate the number of iteration. As in continuous continuation, the subscript i will identify a function or a quantity belonging to the i – th segment where $x_{i-1} \le x \le x_i$.

As can be seen from section 7.1, at each iteration it is necessary to solve the linearized boundary value problem (7.21). Since it is completely identical with the problem (7.37), the algorithm for its solution by the orthogonal shooting method is the same as that described in section 7.2. This algorithm will be used here. Let us introduce the notation

$$L^{(j)} = L(y^{(j)}, p^{(j)}), \quad M^{(j)} = M(y^{(j)}, p^{(j)}), \quad \Phi^{(j)} = \Phi(y^{(j)}, p^{(j)}).$$
$$(7.105)$$

Suppose that for the previous value of λ equal to $\lambda_{k-1} = \lambda_k - \Delta\lambda$ we know the following quantities: first, the vector function $y(x)|_{\lambda=\lambda_{k-1}}$, which is a solution of the nonlinear boundary value problem (7.1) with the corresponding value of the parameter $p = p_{k-1}$, second, the vector function $Y(x)|_{\lambda=\lambda_{k-1}}$, which is a solution of the linearize boundary value problem (7.6) for $\lambda = \lambda_{k-1}$, third, the vector $C_{0(N)}|_{\lambda=\lambda_{k-1}}$, which is a solution of the homogeneous part of the equation (7.71) and corresponds to the vector function $Y(x)|_{\lambda=\lambda_{k-1}}$, fourth, the vector $c_{(N)}|_{\lambda=\lambda_{k-1}}$, which determining the solution of the equation (7.70) and maps of the vector function $y(x)|_{\lambda=\lambda_{k-1}}$ in $^{l+1}$ by virtue of the representation (7.23). We also take into account that $P_{k-1} = dp/d\lambda|_{\lambda=\lambda_{k-1}}$ is the $(l+1)$ – th component of the vector $C_{0(N)}|_{\lambda=\lambda_{k-1}}$ and therefore it is known together with this vector.

The iterative process for $\lambda = \lambda_k$ consists of the following stages.

1. Specification of the starting approximation $(j = 0)$

$$y^{(0)} = y|_{\lambda=\lambda_{k-1}} + \Delta\lambda Y|_{\lambda=\lambda_{k-1}}, \qquad p^{(0)} = p_{k-1} + \Delta\lambda P_{k-1},$$

$$C_{(N)}^{(0)} = C_{(N)}|_{\lambda=\lambda_{k-1}} + \Delta\lambda C_{0(N)}|_{\lambda=\lambda_{k-1}}, \qquad c_{(N)}^{(0)} = c_{(N)}|_{\lambda=\lambda_{k-1}}.$$
$$(7.106)$$

2. Iterative process

j=1

2.1. Forward substitution: construction of general solution matrices and orthogonalization matrices for segments $x_{i-1} \leq x \leq x_i$, $i = \overline{1,N}$

$$\left\{ \frac{dY_{(j)}}{dx} = L^{(j-1)}Y^{(j)} + M^{(j-1)}P^{(j)} + \Phi^{(j-1)}, \right.$$

$$\left. AY^{(j)}(x_0) = a \right\}^{-1} \rightarrow U_{(i)}^{(j)}(x), \qquad i = \overline{1,N}, \qquad (7.107)$$

$$\Omega_{(i)}^{(j)}, \qquad i = \overline{1,N-1}.$$

2.2. Satisfying the boundary conditions for $x = x_N$

$$D^{(j)} = U_{(N)}^{(j)}(x_N), \quad J_{(j)} = BD^{(j)}, \quad d^{(j)} = b - BY_{(N)}^{(l+2)(j)}(x_N),$$

$$\left\{ \begin{bmatrix} J^{(j)} \\ C_{0(N)}^{(j-1)} \end{bmatrix} C_{0(N)}^{(j)} = \begin{bmatrix} 0 \\ 1 \end{bmatrix} \right\}^{-1} \rightarrow C_{0(N)}^{(j)} = \frac{C_{0(N)}^{(j)}}{(C_{0(N)}^{(j)} \cdot C_{0(N)}^{(j)})^{\frac{1}{2}}}, \qquad (7.108)$$

$$\left\{ J^{(j)} C_{(N)}^{(j)} = d^{(j)} \right\}^{-1} \rightarrow g^{(j)}. \qquad (7.109)$$

When solving the equation (7.109) its arbitrary particular solution $g^{(j)}$ should be found.

2.3. Use of the additional condition of the form (1.85)

$$\left\{ C_{(N)}^{(0)} \cdot (C_{(N)}^{(j)} - C_{(N)}^{(j-1)}) = 0 \right\}^{-1} \rightarrow$$

$$\rightarrow a^{(j)} = \frac{C_{(N)}^{(0)} \cdot (c_{(N)}^{(j)} - g_{(N)}^{(j)})}{C_{(N)}^{(0)} \cdot C_{0(N)}^{(j)}}, \qquad (7.110)$$

$$c_{(N)}^{(j)} = a^{(j)} C_{0(N)}^{(j)} + g_{(N)}^{(j)}. \tag{7.111}$$

2.4. Backward substitution.
Consecutive solution of the systems

$$\left\{ c_{(i)}^{(j)} \Omega_{(i)}^{(j)} = c_{(i+1)}^{(j)} \right\} \to c_{(i)}^{(j)}, \quad i = \overline{N-1,1}. \tag{7.112}$$

Construction of the general solutions

$$y^{(j)} = U_{(i)}^{(j)} c_{(i)}^{(j)}, \quad x_{i-1} \le x \le x_i, \quad i = \overline{N,1}. \tag{7.113}$$

2.5. Convergence test
If $[(c_{(N)}^{(j)} - c_{(N)}^{(j-1)}) \cdot (c_{(N)}^{(i)} - c_{(N)}^{(j-1)})]^{-\frac{1}{2}} > \varepsilon > 0$, the computations are repeated beginning with part 2.1 with $j = j+1$. Otherwise, the iterative process can be considered converged and the solution is taken in the form

$$c_{(N)} = c_{(N)}^{(j)}, \quad C_{(N)} = C_{(N)}^{(j)}, \quad y|_{\lambda=\lambda_k} = y^{(j)}, \quad Y|_{\lambda=\lambda_k} = Y^{(j)}. \tag{7.114}$$

These vectors and vector functions are used for specifying the starting approximation in accordance with part 1 for incremented $\lambda_{k+1} = \lambda_k + \Delta\lambda$.

Note that other additional conditions discussed in section 1.5 will only change the computations in part 2.3. Their implementation is not difficult. This is considered in detail in [48]. Some examples of using the algorithm considered here are discussed there in details too. Here we only consider one of the examples.

4.   The example: large deflections of the circle arch

The equations describing the large elastic deflections of the circle arch under uniform normal load p^0 (Fig. 7.1) are considered in detail in [48]. They are

$$x' = (1 + cn)cos\vartheta, \qquad y' = (1 + cn)sin\vartheta,$$

$$\vartheta' = (1 + cn)k, \qquad n' = -(1 + cn)kq, \tag{7.115}$$

$$q' = (1 + cn)(kn - p), \qquad k' = (1 + cn)q.$$

Here $x = x(\beta) = x^0/R$, $y = y(\beta) = y^0/R$ are dimensionless coordinates of the arch axis; β is the length along the nondeformed axis;

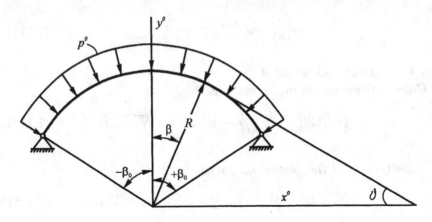

Figure 7.1.

ϑ is the angle between the tangent to the arch axis and axis x; k is a curvature of the deformed axis; n, q are dimensionless longitudinal and shearing forces in the arch; p is dimensionless parameter of the load; $()' = d()/d\beta$; c is a constant determining the flexibility of the arch.

We will consider the boundary conditions corresponding to hinged ends of the arch

$$x(\pm\beta_0) = \pm sin\beta_0, \quad y(\pm\beta_0) = cos\beta_0, \quad k(\pm\beta_0) = -1. \qquad (7.116)$$

The initial undeformed state of the arch is defined by the following relations

$$x_0 = sin\beta, \quad y_0 = cos\beta, \quad \vartheta = -\beta, \quad n_0 = 0,$$

$$q_0 = 0, \quad k_0 = -1, \quad p = 0. \qquad (7.117)$$

As we can see, this relations are the solution of the equations (7.115).

Thus, the equations (7.115) with the boundary conditions (7.116) form nonlinear boundary value problem containing parameter p. In addition, the solution (7.117) of the problem for $p = 0$ is known.

In accordance with the basic idea of the continuation method we assume that the unknown functions x, y, ϑ, n, q, k and the parameter p are functions of the continuation parameter λ the meaning of which has been defined in section 7.1, i.e.,

$$x = x(\beta, \lambda), \quad y = y(\beta, \lambda), \quad \vartheta = \vartheta(\beta, \lambda), \quad n = n(\beta, \lambda),$$

$$q = q(\beta, \lambda), \quad k = k(\beta, \lambda), \quad p = p(\lambda). \qquad (7.118)$$

Then continuation equations (7.5) are represented in the form

$$\frac{dx}{d\lambda} = X, \quad \frac{dy}{d\lambda} = Y, \quad \frac{d\vartheta}{d\lambda} = \Theta,$$

$$\frac{dn}{d\lambda} = N, \quad \frac{dq}{d\lambda} = Q, \quad \frac{dk}{d\lambda} = K, \quad \frac{dp}{d\lambda} = P. \tag{7.119}$$

The linearized boundary value problem for determination of X, Y, Θ, N, Q, K, P is obtained by differentiating the problem (7.115), (7.116) with respect to parameter λ

$$X' = c\cos\vartheta N - (1 + cn)\sin\vartheta\Theta,$$

$$Y' = c\sin\vartheta N + (1 + cn)\cos\vartheta\Theta,$$

$$\Theta' = ckN + (1 + cn)K,$$

$$N' = -ckqN - (1 + cn)(kQ + qK), \tag{7.120}$$

$$Q' = c(kn - p)N + (1 + cn)(kN + nK - P),$$

$$K' = cqN + (1 + cn)Q,$$

$$X(\pm\beta_0) = Y(\pm\beta_0) = K(\pm\beta_0) = 0.$$

The boundary value problem is linear with respect to the unknown functions $X, \ldots, K$ and parameter P, which determine the right hand sides of the continuous equations (7.116).

To reduce the boundary value problem (7.115), (7.119) to the form (7.1) we introduce vector functions $z = [x, y, \vartheta, n, q, k]^T$ and $Z = [X, Y, \Theta, N, Q, K]^T$. After that the problem (7.115), (7.119) can be represented in the form

$$z' = F(z, p), \qquad Az(-\beta_0) = a, \qquad Bz(\beta_0) = b. \tag{7.121}$$

Here the notation $F(z, p) = [F_1(z, p), \ldots, F_6(z, p)]^T$ has been introduced for nonlinear vector function corresponding to the equations (7.115). The rectangular matrices A, B of size 3×6 and the three − dimensional vectors a, b are determined by the conditions (7.116) and can be written as

$$A = B = \begin{bmatrix} 1 & 0 & 0 & 0 & 0 & 0 \\ 0 & 1 & 0 & 0 & 0 & 0 \\ 0 & 0 & 0 & 0 & 0 & 1 \end{bmatrix}, \quad a = \begin{bmatrix} -\sin\beta_0 \\ \cos\beta_0 \\ -1 \end{bmatrix}, \quad b = \begin{bmatrix} \sin\beta_0 \\ \cos\beta_0 \\ -1 \end{bmatrix}. \tag{7.122}$$

In this notations the continuation equations (7.119) are represented in the form

$$\frac{dz}{d\lambda} = Z, \qquad \frac{dp}{d\lambda} = P. \qquad (7.123)$$

We have the following initial conditions

$$z(a) = z_{(0)} = [sin\beta, cos\beta, -\beta, 0, 0, -1]^{T}, \quad p(0) = p_0 = 0. \qquad (7.124)$$

The boundary value problem (7.120) becomes

$$Z' = L(z,p)Z + PM(z,p), \qquad AZ(-\beta) = 0, \quad AZ(\beta) = 0. \qquad (7.125)$$

Here $L(z,p)$ and $M(z,p)$ are a matrix and a vector whose components for the nonlinear vector function $F(z,p)$ of the right hand sides of equations (7.115) are defined by the following relations:

$$L = [L_{ij}] = \frac{\partial F}{\partial z} = \left[\frac{\partial F_i}{\partial z_j}\right], \qquad M = [M_i]^T = \left[\frac{\partial F_i}{\partial p}\right]^T,$$

$$i,j = \overline{1,6}. \qquad (7.126)$$

The expended matrix form of the equations (7.125) is given as

$$
\begin{bmatrix} X \\ Y \\ \Theta \\ N \\ Q \\ K \end{bmatrix}'
=
\begin{bmatrix}
0 & 0 & L_{13} & L_{14} & 0 & 0 \\
0 & 0 & L_{23} & L_{24} & 0 & 0 \\
0 & 0 & 0 & ck & 0 & L_{36} \\
0 & 0 & 0 & L_{44} & L_{45} & L_{46} \\
0 & 0 & 0 & L_{54} & 0 & L_{56} \\
0 & 0 & 0 & cq & L_{65} & 0
\end{bmatrix}
\begin{bmatrix} X \\ Y \\ \Theta \\ N \\ Q \\ K \end{bmatrix}
+ P
\begin{bmatrix} 0 \\ 0 \\ 0 \\ 0 \\ M_5 \\ 0 \end{bmatrix}.
$$

Here $L_{13} = -(1 + cn)sin\vartheta$, $L_{14} = ccos\vartheta$, $L_{23} = (1 + cn)cos\vartheta$, $L_{24} = csin\vartheta$, $L_{36} = 1 + cn$, $L_{44} = -ckq$, $L_{45} = -(1 + cn)k$, $L_{46} = -(1 + cn)q$, $L_{54} = (1 + 2cn)k - cp$, $L_{56} = (1 + cn)n$, $L_{65} = 1 + cn$, $M_5 = 1 + cn$.

Since simple Euler method leads to accumulating of considerable error we adopted the algorithm of the modified Euler method (7.94) – (7.104) as a basic method for integrating the continuation equations (7.123). To eliminate the error, the modified Euler method was combined with the discrete continuation procedure for which we applied the algorithm in the form (7.106) – (7.114) choosing the coefficient $a^{(j)}$ according to (7.110). For the sake of simplicity, we use capital letters in the equations for quasi linearization (7.21) to denote the unknown functions of the current $(j+1)$ – th approximation and small letters for the known functions of the previous approximation for $\lambda = \lambda_k$

$$Z = z_{(k)}^{(j+1)}, \quad P = p_k^{(j+1)}, \quad z = z_{(k)}^{(j)}, \quad p = p_k^{(j)}. \tag{7.127}$$

The boundary value problem (7.21) becomes then

$$Z' = L(z,p)Z + PM(z,p) + [-L(z,p)z - pM(z,p) + F(z,p)],$$
$$AZ(-\beta_0) = a, \quad BZ(\beta_0) = b. \tag{7.128}$$

Comparing these equations (7.128) with the equations (7.125), we note that they differ only by the presence of bracketed terms. In expanded form, the boundary value problem (7.128) is

$$X' = c\cos\vartheta N - (1+cn)\sin\vartheta\Theta+$$
$$+[-c\cos\vartheta n + (1+cn)\sin\vartheta\vartheta + (1+cn)\cos\vartheta],$$

$$Y' = c\sin\vartheta N + (1+cn)\cos\vartheta\Theta+$$
$$+[-c\sin\vartheta n - (1+cn)\cos\vartheta\vartheta + (1+cn)\sin\vartheta],$$

$$\Theta' = ckN + (1+cn)K+ \tag{7.129}$$
$$+[-ckn - (1+cn)k + (1+cn)k],$$

$$N' = -ckqN - (1+cn)(kQ+qK)+$$
$$+[ckqn + (1+cn)(kq+qk) - (1+cn)kq],$$

$$Q' = c(kn - p)N + (1 + cn)(kN + nK) - (1 + cn)P+$$

$$+[-c(kn - p)n - (1 + cn)(kn + nk) + (1 + cn)kn],$$

$$K' = cqN + (1 + cn)Q+ \qquad \qquad (7.130)$$

$$+[-cqn - (1 + cn)q + (1 + cn)q],$$

$$X(\pm\beta_0) = \pm sin\beta_0, \quad Y(\pm\beta_0) = \pm cos\beta_0, \quad K(\pm\beta_0) = 1.$$

We have tried to write these equations in a form which emphasizes the similarity of the terms on the right-hand sides which are written one under another and underlined. The same similarity is seen in the generalized form of (7.128), where similar terms are underlined in a similar manner. This enables one to use the same procedures for both continuous and discrete continuation in developing software.

In the practical implementation of these algorithms [105] for integrating linear equations (7.120), (7.129), (7.130) we used the Runge - Kutta method for obtaining their fundamental solutions in the forward and backward substitutions of shooting method. The test computations have shown that sufficient accuracy is achieved by dividing the arch with $\beta_0 = 22.5^o$ into 30 segments and that with $\beta_0 = 90^o$ into 100 segments.

In developing the algorithms and softwares, a number of test computations were carried out to investigate the effect of step size with respect to λ on the error accumulation and the effectiveness of combining continuous and discrete continuations. Fig. 7.2 shows the relations between the dimensionless load p and the fractional deflection $W = w/R$ of the middle point ($\beta = 0$) of the arch during its symmetric deformation. The computations were performed for the arch with parameter $c = 10^{-4}$ and angle $\beta = 45^o$.

The dashed curve 1 in Fig. 7.2 corresponds to integrating the continuation equations by the modified Euler method with the step $\Delta\lambda$ which in the initial stage of the deformation for small P corresponded to increment of the fractional deflection $W = 0.005$. The dot-and-dash curve 2 corresponds to the same method but with the step $W = 0.0025$. The solid line 3 was obtained by combining two steps $W = 0.005$ of the modified Euler method with one step of the implicit discrete continuation scheme. This curve corresponds practically to the exact solution of the problem. As can be seen from Fig. 7.2, the modified Euler method results in substantial error accumulation especially for the ranges of parameter, where the solution changes considerably. At the same time

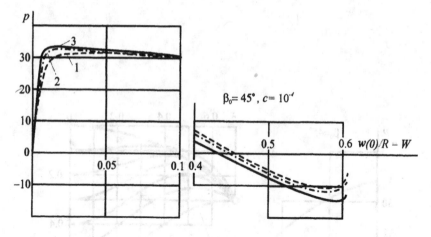

Figure 7.2.

computational cost of obtaining curves 2 and 3 is practically the same (for the curve 3 it was somewhat smaller). It allows to recommend the combination of continuous and discrete continuation for computations.

Fig. 7.3 shows the deformation curves $p(W)$ obtained for the arch with $\beta = 45^o$. Both symmetric and asymmetric shapes of deformation that result from the buckling of the arch were considered.

To obtain postbuckling forms of deformation, the perturbation of the load were introduced close to bifurcation points. For example, for symmetric shape the load was taken in the form

$$p(\beta) = p[1 + 0.01sin(\pi(\beta - \beta_0)/2\beta_0)],$$

and for asymmetric shape in the form

$$p(\beta) = p[1 + 0.01cos(\pi(\beta - \beta_0)/2\beta_0)].$$

Introducing of the such perturbation allows by-pass the bifurcation points along perturbed solutions.

Another examples of implementation of the considered algorithms are given in chapter 4 of the book [48].

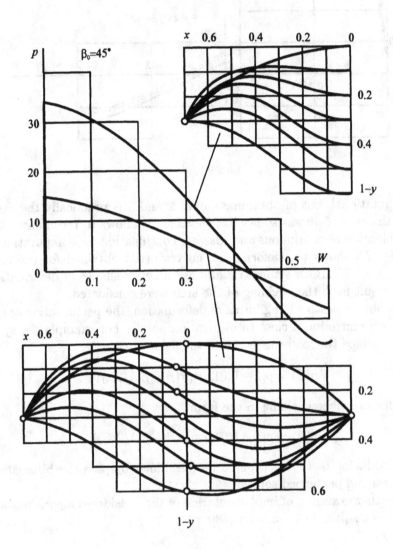

Figure 7.3.

Chapter 8

CONTINUATION OF THE SOLUTION NEAR SINGULAR POINTS

The continuation methods developed in chapter 1 treat the unknowns and the parameter the same way. They use a universal continuation algorithm at regular and limit points of the solution set of nonlinear system equations. From standpoint of the continuation algorithm, it is, therefore, unnecessary to introduce the concept of a limit point. Further to the discussion in chapter 1, here the primary attention is given to analysis of the behavior of the solution in the neighborhood of essentially singular points, i.e., the points where the augmented Jacobian matrix J is singular. As a basic method of analysis we adopt the method of Taylor series in the neighborhood of singular point. This enables us to construct the bifurcation equation, and , analyzing it, to find all brunches of the solution. The complexity of the analysis depends on the degree of singularity of matrix J. We shall consider the case of simple singularity of matrix J ($rank(J) = n - 1$), which is the most important case for applications, and also a more complicated case of its double singularity ($rank(J) = n - 2$).

1. Classification of singular points

As in Chapter 1, the problem of continuation of the solution for the system of equations (1.1) is considered in an $(n + 1)$-dimensional Euclidean $^{n+1}$ in which a vector $x = (x_1, \ldots, x_n, x_{n+1} = p)^T$ is introduced. The problem then reduces to continuing the solution of the system of equations

$$F_i(x) = 0, \qquad i = \overline{1, n}. \tag{8.1}$$

Let the functions $F_i(x)$ be analytic; the components of the vector x may then be regarded as functions of the continuation parameter λ

$$x_i = x_i(\lambda), \qquad i = \overline{1, n+1}. \tag{8.2}$$

Suppose that at a certain point, for which the parameter λ is taken to be zero, the solution $x(0) = x_{(0)}$ is known. The behavior of the solution $x(\lambda)$ in the neighborhood of this point is determined by expansion in a Taylor power series

$$x(\lambda) = x_{(0)} + x'_{(0)}\lambda + \frac{1}{2!}x''_{(0)}\lambda^2 + \frac{1}{3!}x'''_{(0)}\lambda^3 + \ldots \tag{8.3}$$

The notation is

$$x'_{(0)} = dx/d\lambda|_{\lambda=0}, \quad x''_{(0)} = d^2x/d\lambda^2|_{\lambda=0}, \ldots \tag{8.4}$$

The equation for determining $x'_{(0)}, \quad x''_{(0)}, \ldots$ is obtained by successively differentiating Eq. (8.1) with respect to λ

$$F^0_{i,j}x'_{(0)j} = 0, \qquad i = \overline{1, n}. \tag{8.5}$$

$$F^0_{i,j}x''_{(0)j} + F^0_{i,jk}x'_{(0)j}x'_{(0)k} = 0, \quad i = \overline{1, n}. \tag{8.6}$$

$$F^0_{i,j}x'''_{(0)j} + 3F^0_{i,jk}x''_{(0)j}x'_{(0)k} + F^0_{i,jkl}x'_{(0)j}x'_{(0)k}x'_{(0)l} = 0,$$
$$i = \overline{1, n}. \tag{8.7}$$

. . . .

In these equations the summation sign is assuming that the repeated indexes must be summed from 1 to $n + 1$, and the following notation is used:

$$F^0_{i,j} = \partial F_i/\partial x_j|_{\lambda=0}, \quad F^0_{i,jk} = \partial^2 F_i/\partial x_j\partial x_k|_{\lambda=0}, \ldots \tag{8.8}$$

The sequence of systems of equations (8.5), (8.6),... is recurrent, and in each of the systems the coefficients of the unknowns form the augmented Jacobian matrix $J^0 = J(x_{(0)}) = [F^0_{i,j}], \quad i = \overline{1, n}, \quad j = \overline{1, n+1}$.

Note that the first of these systems, (8.5), is homogeneous, and all the rest are nonhomogeneous.

At regular and limit points of the solution set K of system (8.1) in $^{n+1}$, the matrix J is nonsingular, i.e., $rank(J) = n$. Therefore, the solution of the homogeneous system (8.5), as previously shown in Section 1.5, belongs to a one-dimensional subspace $A^1 \subset {}^{n+1}$. According to (8.5), the subspace A^1 is orthogonal to an n-dimensional subspace $P^n \subset {}^{n+1}$ which is determined by the basis of n linearly independent row vectors of the matrix J. Let $a^{(1)} = (a_1^{(1)}, \ldots, a_{n+1}^{(1)})$ be the unit vector of the subspace A^1. The solution of system (8.5) is then represented as

$$x'_{(0)} = ca^{(1)}. \tag{8.9}$$

Here c is an arbitrary constant coefficient. Since the vector $x'_{(0)}$ only changes its direction as λ varies and the length defined by relation (8.9) remains unchanged, the vectors $x''_{(0)}, x'''_{(0)}, \ldots,$, characterizing the change in the direction of the vector $x'_{(0)}$, must be orthogonal to it and so must belong to the subspace P^n, which is an orthogonal complement of the subspace A^1 in $^{n+1}$. Thus, as in Section 1.5,

$$^{n+1} = P^n \oplus A^1 \tag{8.10}$$

From (8.9) it follows immediately that for $c = 1$ the parameter λ is the differential of length of the solution curve K of system (8.1) in $^{n+1}$. Indeed, since $a^{(1)}$ is the unit vector, i.e., $(a^{(1)}, a^{(1)}) = 1$, it follows that

$$(x'_{(0)}, x'_{(0)}) = \sum_{j=1}^{l+1} (dx_j/d\lambda|_{\lambda=0})^2 = c^2. \tag{8.11}$$

Hence

$$\left(\sum_{j=1}^{l+1} (dx_j)^2 \right)^{1/2} = cd\lambda. \tag{8.12}$$

Since $x \in K$, the left-hand side of this relation is the differential of length of the curve K. Therefore, for $c = 1$ the right-hand side $d\lambda$ is also the differential of length of K.

Consider, now, a point $x_{(0)}$ at which $rank(J(x_{(0)})) = r < n$. This means that among n rows of the matrix $J^0 = J(x_{(0)})$ only r rows are linearly independent. For accuracy, we assume that the first r rows of the matrix J^0 are linearly independent. Any other case can always be reduced to this one by an obvious renumbering of the equations in system (8.1).

Let us divide Eqs. (8.1) into two groups

$$F_i(x) = 0, \qquad i = \overline{1, r}, \tag{8.13}$$

$$F_j(x) = 0, \qquad j = \overline{r+1, n}, \tag{8.14}$$

To simplify subsequent computations, suppose that the origin of coordinates in the space $^{n+1}$ is placed at the point $x_{(0)}$, in the neighborhood of which the behavior of solution is being investigated, i.e., $x_{(0)} = 0$. This can also be achieved by introducing a new unknown y in place of x, such that $y = x - x_{(0)}$. Thus, we assume that

$$F_i(0) = 0, \qquad i = \overline{1, n}. \tag{8.15}$$

Differentiate Eqs. (8.13) with respect to parameter λ. As a result, we obtain at the point $x_{(0)} = 0$

$$F^0_{i,j} x'_{(0)j} = 0, \quad i = \overline{1, r}, \quad j = \overline{1, n+1}. \tag{8.16}$$

Let the Jacobian matrix of system (8.13) be denoted by $J_r = [F_{i,j}]$, $(i = \overline{1, r}, \ j = \overline{1, n})$. It has r rows and $n + 1$ columns. By construction, its rows are linearly independent at the point $x_{(0)} = 0$, and therefore its rank at this point is r

$$rank(J_r(0)) = r. \tag{8.17}$$

Represent $^{n+1}$ as a direct sum of two subspaces

$$^{n+1} = P^r \oplus A^d, \qquad d = n + 1 - r. \tag{8.18}$$

The former, P^r, is an r-dimensional subspace in $^{n+1}$ whose basis is formed by the row vectors of the matrix $J^0_r = J_r(0)$, and the latter, A^d, is an orthogonal complement of P^r in $^{n+1}$.

Let $p^{(i)} \in P^r$, $(i = \overline{1, r})$ and $a^{(j)} \in A^d$, $(j = \overline{1, d})$ be orthogonal bases in P^r and A^d, respectively. Henceforth, $p^{(i)}$ is assumed to be a basis constructed of the rows of the matrix J_r^0 by Gram-Schmidt process (Section 7.2). By construction, the following relations hold for the bases thus introduced:

$$(p^{(i)}, p^{(j)}) = \delta_{ij}, \qquad i, j = \overline{1, r}, \tag{8.19}$$

$$(a^{(i)}, a^{(j)}) = \delta_{ij}, \qquad i, j = \overline{1, d}, \tag{8.20}$$

$$(p^{(i)}, a^{(j)}) = 0, \qquad i = \overline{1, r}, \quad j = \overline{1, d}, \tag{8.21}$$

$$\delta_{ij} = \begin{cases} 1 & \text{if } i = j, \\ 0 & \text{if } i \neq j. \end{cases}$$

Here δ_{ij} is the Kronecker symbol. Taken together, the bases $p^{(i)}$ and $a^{(i)}$ from a basis in $^{n+1}$, and so it is clear that every vector $x \in {}^{n+1}$ can be represented uniquely as a decomposition along the basis vectors $p^{(i)}$ and $a^{(i)}$

$$x = \sum_{i=1}^{r} \rho_i p^{(i)} + \sum_{j=1}^{d} \alpha_j a^{(j)}. \tag{8.22}$$

Conversely, for each decomposition of the form (8.22) there corresponds a unique vector $x \in {}^{n+1}$. If, in addition, $x \equiv 0$, then $\rho_i = 0$ $(i = \overline{1, r})$ and $\alpha_j = 0$ $(j = \overline{1, d})$. Thus, relation (8.22) determines a one-to-one correspondence between the components of the vector x and the coefficients of its decomposition ρ_i, α_j. This enables one to make a change of the variables in Eqs. (8.13), (8.14) by means of (8.22). These equations then become

$$F_i(\rho_1, \ldots, \rho_r, \alpha_1, \ldots, \alpha_d) = 0, \quad i = \overline{1, r}, \tag{8.23}$$

$$F_j(\rho_1, \ldots, \rho_r, \alpha_1, \ldots, \alpha_d) = 0, \quad j = \overline{r+1, n}. \tag{8.24}$$

We also transform to the unknowns ρ_i, α_j in the continuation equations (8.16). They take the form

$$\sum_{j=1}^{n+1} F_{i,j}^0 \left(\sum_{k=1}^{r} \rho_k' p^{(k)} + \sum_{l=1}^{d} \alpha_l' a^{(l)} \right) = 0, \quad i = \overline{1, r}. \tag{8.25}$$

By applying the Gram-Schmidt process we represent the matrix of this equation $J_r^0 = [F_{i,j}]$ $(i = \overline{1,r},\ j = \overline{1,n+1})$ in the form of the product of the orthogonalization matrix Ω of order r and the orthogonal matrix P of size $r \times (n+1)$ whose rows are the vectors $p^{(i)}$ $(i = \overline{1,r})$ of the orthonormal basis of the subspace $P^r \subset {}^{n+1}$

$$ J_r^0 = \Omega P, \qquad P = \begin{bmatrix} p^{(1)T} \\ \vdots \\ p^{(r)T} \end{bmatrix}. \qquad (8.26) $$

If, taking into account this representation, we multiply system (8.25) on the left by the matrix Ω^{-1}, it becomes

$$ P\left(\sum_{k=1}^{r} \rho_k' p^{(k)} + \sum_{l=1}^{d} \alpha_l' a^{(l)} \right) = 0. \qquad (8.27) $$

By writing out these equations, with (8.26), and using relations (8.19) – (8.21), we obtain

$$ E_r \rho' = 0, \qquad \rho' = (\rho_1', \ldots, \rho_r')^T. \qquad (8.28) $$

Here E_r is a square identity matrix of order r.

The result obtained allows two conclusions to be drawn. First, $\rho_k' = 0$, $(k = \overline{1,r})$, which once more confirms the fact that the nonzero solutions x' of Eq. (8.16) belong to the subspace A^d. Secondly, the determinant of system (8.28) is equal to unity and, apart from a nonzero constant factor $det(\Omega^{-1})$, coincides with the Jacobian of Eqs. (8.23) with respect to the variables ρ_k $(k = \overline{1,r})$. By use of the implicit function theorem, in a small neighborhood of the point $x = 0$ in question, the variables ρ_k $(k = \overline{1,r})$ can then be obtained from Eqs. (8.23) as functions of the variables α_l $(l = \overline{1,d})$.

$$ \rho_k = \rho_k(\alpha_1, \ldots, \alpha_d), \qquad k = \overline{1,r}, \qquad (8.29) $$

The functions ρ_k are single valued, continuous and differentiable. Substituting expressions (8.29) in Eqs. (8.24) leads to the bifurcation equations

$$ F_j(\rho_1(\alpha_1, \ldots, \alpha_d), \ldots, \rho_r(\alpha_1, \ldots, \alpha_d), \alpha_1, \ldots, \alpha_d) = 0, $$

$$ j = \overline{r+1, n}. \qquad (8.30) $$

These equations determine both the number of solution branches and their behavior in the neighborhood of the point under investigation.

Since the bifurcation equations (8.30) are constructed so that their Jacobian is zero, they may have a nonunique solution. For each solution of these equations as a function of the parameter λ

$$\alpha_l^{(t)} = \alpha_l^{(t)}(\lambda), \qquad l = \overline{1, d}, \tag{8.31}$$

from Eqs. (8.30) or, alternatively, from Eqs. (8.24) we obtain

$$\rho_k^{(t)} = \rho_k^{(t)}(\alpha_1^{(t)}, \ldots, \alpha_d^{(t)}), \qquad k = \overline{1, r}, \tag{8.32}$$

It is then possible to construct one of the solution branches $x^{(t)}(\lambda)$ whose behavior in a small neighborhood of the point being considered is determined by decomposition (8.22).

It should be noted that the derivation of the bifurcation equations is rather complicated, and they can be written explicitly only in exceptional cases. Their solution in explicit analytic form can be found even less often. Therefore, the methods which avoid direct use of bifurcation equation in the form (8.30) but instead solve a problem on the basis of simpler relations are fundamentally important.

2. The simplest form of bifurcation equations

Let us transform system (8.1)so that its linear part assumes the simplest form. For further discussion, it is convenient to use the following notation for the Jacobian matrix (see.e.g., [40]):

$$J = \frac{\partial F}{\partial x} = \frac{\partial(F_1, \ldots, F_n)}{\partial(x_1, \ldots, x_{n+1})}. \tag{8.33}$$

As before, the rows of the matrix $[F_{i,1}, \ldots, F_{i,n+1}]$ $(i = \overline{1, n})$, are regarded as vectors in the space $^{n+1}$ and are denoted by $f^{(i)}$. Then

$$f^{(i)} = [F_{i,1}, \ldots, F_{i,n+1}]^T. \tag{8.34}$$

As in Section 8.1, let the solution of system (8.1) be investigated in the neighborhood of the point $x = 0$,and assume that at this point $rank(J^0) = r < n$. Henceforth, the superscript '0' of a function will be used to indicate the value of this function at the point $x = 0$. Let the first r rows of the Jacobian matrix J^0 again be linearly independent.

The last $n - r = d - 1$ rows of this matrix are then linear combinations of the first r rows, i.e.,

$$f^{0(r+i)} = \sum_{k=1}^{r} d_{ik} f^{0(k)}, \qquad i = \overline{1, d-1}. \tag{8.35}$$

As in Section 8.1, we divide system (8.1) into two groups of equations, namely (8.13), (8.14). Denote the Jacobian matrices of these groups of equations by $J^{(1)}$ and $J^{(2)}$

$$J^{(1)} = \frac{\partial(F_1, \ldots, F_r)}{\partial(x_1, \ldots, x_{n+1})}, \quad J^{(2)} = \frac{\partial(F_{r+1}, \ldots, F_n)}{\partial(x_1, \ldots, x_{n+1})}. \tag{8.36}$$

It is clear that

$$J = \begin{bmatrix} J^{(1)} \\ J^{(2)} \end{bmatrix}. \tag{8.37}$$

If we introduce a matrix $D = [d_{ik}]$ $(i = \overline{1, d-1}, \ k = \overline{1, r})$, representation (8.35) in matrix form becomes

$$J^{0(2)} = D J^{0(1)}. \tag{8.38}$$

As in Section 8.1, we represent the space $^{n+1}$ as a direct sum of two orthogonal subspaces

$$^{n+1} = P^r \oplus A^d, \qquad d = n + 1 - r. \tag{8.39}$$

Here the r-dimensional subspace P^r is determined by the basis formed by the row vectors of the matrix $J^{0(1)}$ and A^d is its orthogonal complement in $^{n+1}$.

Introduce in P^r an orthonormal basis $p^{(i)}$ $(i = \overline{1, r})$, constructed of the row vectors of the matrix $J^{0(1)}$ by the Gram-Schmidt process. Then

$$J^{0(1)} = \Omega P. \tag{8.40}$$

Here the orthogonal matrix P is of size $r \times (n + 1)$ and its rows are the vectors $p^{(i)}$ of the orthonormal basis in P^r

$$P = \begin{bmatrix} p^{(1)T} \\ \cdots \\ p^{(r)T} \end{bmatrix}, \tag{8.41}$$

Ω is the left triangular orthogonalization matrix.

Introduce in the subspace A^d an orthonormal basis $a(i)$ $(i = \overline{1,d})$ and a matrix A of size $d \times (n+1)$ whose rows are the vectors $a^{(i)}$ $(i = \overline{1,d})$,

$$
A = \begin{bmatrix} a^{(1)T} \\ \cdots \\ a^{(d)T} \end{bmatrix}.
\tag{8.42}
$$

Transform the original system of equations (8.1) as follows. Multiply Eq. (8.13) on the left by the matrix Ω^{-1}. The resulting system of equations is

$$
\Omega^{-1} \begin{bmatrix} F_1(x) \\ \cdots \\ F_r(x) \end{bmatrix} = \begin{bmatrix} U_1(x) \\ \cdots \\ U_r(x) \end{bmatrix} = U(x) = 0.
\tag{8.43}
$$

By virtue of the linearity of the transformation and the nonsingularity of the matrix Ω^{-1} Eqs. (8.13) and (8.43) are equivalent in the sense that all solutions x of Eqs. (8.13) are the solutions of (8.43), and vice versa, i.e., the solution sets of Eqs. (8.13) and (8.43) coincide. The Jacobian matrix of system (8.43) as $x = 0$ is the orthogonal matrix P. Indeed, from (8.43) and (8.40) it follows that

$$
\frac{\partial(U_1,\ldots,U_r)^0}{\partial(x_1,\ldots,x_{n+1})} = \frac{\partial U^0}{\partial x} = \Omega^{-1} \frac{\partial(F_1,\ldots,F_r)^0}{\partial(x_1,\ldots,x_{n+1})} = \Omega^{-1}\Omega P = P.
\tag{8.44}
$$

On the basis of Eqs. (8.14) we derive the following equations:

$$
F_{r+i}(x) - \sum_{k=1}^{r} d_{ik} F_k(x) = V_i(x) = 0, \quad i = \overline{1, d-1}.
\tag{8.45}
$$

which, in matrix form, becomes

$$
\begin{bmatrix} F_{r+1}(x) \\ \cdots \\ F_n(x) \end{bmatrix} - D \begin{bmatrix} F_1(x) \\ \cdots \\ F_r(x) \end{bmatrix} = \begin{bmatrix} V_1(x) \\ \cdots \\ V_{d-1}(x) \end{bmatrix} = V(x) = 0.
\tag{8.46}
$$

The Jacobian matrix of this system as $x = 0$ vanishes. Indeed, from (8.46), by (8.38), we obtain

$$
\frac{\partial(V_1,\ldots,V_{d-1})^0}{\partial(x_1,\ldots,x_{n+1})} = \frac{\partial V^0}{\partial x} = J^{0(2)} - D J^{0(1)} = 0.
\tag{8.47}
$$

It is easy to see that the solution sets of system (8.1) and systems (8.43), (8.46) coincide. Of course, we exclude the case when at least one of the functions V_i $(i = \overline{1, d-1})$ is identically zero. This case merely implies a reduction in the number of equations in system (8.1).

As a result of transformations, system (8.1) reduces to an equivalent system of n equations

$$U_i(x) = 0, \qquad i = \overline{1, r}, \tag{8.48}$$

$$V_j(x) = 0, \qquad j = \overline{1, d-1}. \tag{8.49}$$

By construction, the Jacobian matrix of this system for $x = 0$ is of the form

$$\frac{\partial(U, V)^0}{\partial x} = \left[\begin{array}{c} \partial U^0/\partial x \\ \partial V^0/\partial x \end{array} \right] = \left[\begin{array}{c} P \\ 0 \end{array} \right]. \tag{8.50}$$

Taken together, the orthonormal bases $p^{(i)}$ $(i = \overline{1, r})$ and $a^{(j)}$ $(j = \overline{1, d})$ form, by (8.39), an orthonormal basis in $^{n+1}$. The desired solution can therefore be represented as an expansion in these orthonormal bases

$$x = \sum_{i=1}^{r} \rho_i p^{(i)} + \sum_{j=1}^{d} \alpha_j a^{(j)} = \rho P + \alpha A =$$

$$= [\rho, \alpha] \left[\begin{array}{c} P \\ A \end{array} \right] = [P^T A^T] \left[\begin{array}{c} \rho^T \\ \alpha^T \end{array} \right]. \tag{8.51}$$

Here $\rho = (\rho_1, \ldots, \rho_r)^T$ and $\alpha = (\alpha_1, \ldots, \alpha_r)^T$ may be regarded as vectors in Euclidean spaces $\mathcal{P}^r$ and $\mathcal{A}^d$ of dimension r and d, respectively, such that there is a one-to-one correspondence with the subspaces P^r and A^d set up by relation (8.51).

Let us change in Eqs. (8.48), (8.49) from the unknown x to the unknowns ρ and α. The result is

$$U_i(\rho_1, \ldots, \rho_r, \alpha_1, \ldots, \alpha_d) = 0, \quad i = \overline{1, r}, \tag{8.52}$$

$$V_j(\rho_1, \ldots, \rho_r, \alpha_1, \ldots, \alpha_d) = 0, \quad j = \overline{1, d-1}. \tag{8.53}$$

Or, in vector form,

$$\left[\begin{array}{c} U(\rho, \alpha) \\ V(\rho, \alpha) \end{array} \right] = 0. \tag{8.54}$$

The Jacobian matrix of this system with respect to the variables ρ, α for $x = 0$ is of the form

$$\frac{\partial(U, V)^0}{\partial(\rho, \alpha)} = \frac{\partial(U, V)^0}{\partial x}\frac{\partial x}{\partial(\rho, \alpha)} =$$

$$= \begin{bmatrix} P \\ 0 \end{bmatrix}[P^T\ A^T] = \begin{bmatrix} PP^T & PA^T \\ 0 & 0 \end{bmatrix} = \begin{bmatrix} E & 0 \\ 0 & 0 \end{bmatrix}. \tag{8.55}$$

The product PP^T is equal to the identify matrix E of order r since P is an orthogonal matrix; $PA^T = 0$ by (8.21).

Thus, as a result of transformations (8.43), (8.45) and (8.51), we have changed from system (8.1) with the unknown x to system (8.54)with the unknowns ρ, α. Again the transformations are such that there is a one-to-one correspondence between the set of solutions of these systems. But the Jacobian matrix of system (8.54) at the singular point $\rho = \alpha = 0$ ($x = 0$) has the simplest form (8.55). Taking this into account, by the implicit function theorem, ρ may be expressed in terms of α from Eqs. (8.52)

$$\rho = \rho(\alpha). \tag{8.56}$$

These expressions cannot include relations linear in α. Substituting (8.56) in (8.53), the bifurcation equations are obtained as

$$V(\rho(\alpha), \alpha) = 0. \tag{8.57}$$

By the structure of the Jacobian matrix (8.55), these functions cannot contain relations linear in α.

As has been noted previously, the bifurcation equation (8.57) in explicit form can only be derived in exceptional cases. The methods for its approximation and analysis has been extensive considered in literature [117, 62], etc. going back to Lyapunov [77] and Schmidt [103]. We shall restrict ourselves here to a method based on the analysis of Taylor series expansions of the form (8.3), (8.5) – (8.7) in the neighborhood of a singular point. By representation (8.51), the behavior of the solution $x(\lambda)$ in close neighborhood of a singular point $\lambda = 0$ is determined by Taylor series expansion in powers of λ

$$x(\lambda) = P^T\left(\rho'_{(0)}\lambda + \frac{1}{2!}\rho''_{(0)}\lambda^2 + \ldots\right) +$$

$$+ A^T\left(\alpha'_{(0)}\lambda + \frac{1}{2!}\alpha''_{(0)}\lambda^2 + \ldots\right), \tag{8.58}$$

$$\rho'_{(0)} = d\rho/d\lambda|_{\lambda=0}, \quad \rho''_{(0)} = d^2\rho/d\lambda^2|_{\lambda=0}, \ldots$$

The equations for determining $\rho'_{(0)}, \rho''_{(0)}, \ldots, \alpha'_{(0)}, \alpha''_{(0)}, \ldots$ are obtained by differentiating Eqs. (8.48), (8.49) with respect to λ. After formal differentiation we have from (8.48)

$$U^0_{,\rho}\rho'_{(0)} + U^0_{,\alpha}\alpha'_{(0)} = 0,$$

$$U^0_{,\rho}\rho''_{(0)} + U^0_{,\alpha}\alpha''_{(0)} + U^0_{,\rho\rho}\rho'_{(0)}\rho'_{(0)} +$$

$$+2U^0_{,\rho\alpha}\rho'_{(0)}\alpha'_{(0)} + U^0_{,\alpha\alpha}\alpha'_{(0)}\alpha'_{(0)} = 0,$$

$$U^0_{,\rho}\rho'''_{(0)} + U^0_{,\alpha}\alpha'''_{(0)} + 3U^0_{,\rho\rho}\rho''_{(0)}\rho'_{(0)} + 3U^0_{,\rho\alpha}\rho''_{(0)}\alpha'_{(0)} + \qquad (8.59)$$

$$+3U^0_{,\alpha\rho}\rho'_{(0)}\alpha''_{(0)} + 3U^0_{,\alpha\alpha}\alpha''_{(0)}\alpha'_{(0)} + U^0_{,\rho\rho\rho}\rho'_{(0)}\rho'_{(0)}\rho'_{(0)} +$$

$$+3U^0_{,\rho\rho\alpha}\rho'_{(0)}\rho'_{(0)}\alpha'_{(0)} + 3U^0_{,\rho\alpha\alpha}\rho'_{(0)}\alpha'_{(0)}\alpha'_{(0)} +$$

$$+U^0_{,\alpha\alpha\alpha}\alpha'_{(0)}\alpha'_{(0)}\alpha'_{(0)} = 0.$$

The meaning of the notation adopted in these equations becomes clear from a comparison of the second equation with its extended form

$$\frac{\partial U^0_i}{\partial \rho_j}\frac{d^2\rho_{(0)j}}{d\lambda^2} + \frac{\partial U^0_i}{\partial \alpha_j}\frac{d^2\alpha_{(0)j}}{d\lambda^2} + \frac{\partial^2 U^0_i}{\partial \rho_j \partial \rho_k}\frac{d\rho_{(0)j}}{d\lambda}\frac{d\rho_{(0)k}}{d\lambda} +$$

$$+2\frac{\partial^2 U^0_i}{\partial \rho_j \partial \alpha_k}\frac{d\rho_{(0)j}}{d\lambda}\frac{d\alpha_{(0)k}}{d\lambda} + \frac{\partial^2 U^0_i}{\partial \alpha_j \partial \alpha_k}\frac{d\alpha_{(0)j}}{d\lambda}\frac{d\alpha_{(0)k}}{d\lambda} = 0, \quad i = \overline{1, r}.$$

Here the repeated indices of ρ are summed from 1 to r, and those of α from 1 to d.

From (8.55) it follows that

$$U^0_{,\rho} = E, \qquad U^0_{,\alpha} = 0. \qquad (8.60)$$

From the first equation of (8.59) we then obtain the result which is already known to us from Section 8.1, namely

$$\rho'_{(0)} = 0, \qquad (8.61)$$

In accordance with representation (8.51), this indicates that vector $x' = dx/d\lambda$ belongs to the subspace A^d.

Expressions (8.60) and (8.61) make it possible to simplify the second and subsequent equations in (8.59). They become

$$\rho''_{(0)} + U^0_{,\alpha\alpha}\alpha'_{(0)}\alpha'_{(0)} = 0,$$

$$\rho'''_{(0)} + 3U^0_{,\rho\alpha}\rho''_{(0)}\alpha'_{(0)} + 3U^0_{,\alpha\alpha}\alpha''_{(0)}\alpha'_{(0)} + \qquad (8.62)$$

$$+U^0_{,\alpha\alpha\alpha}\alpha'_{(0)}\alpha'_{(0)}\alpha'_{(0)} = 0.$$

This sequence of equations enables one to determine $\rho''_{(0)}, \rho'''_{(0)}, \ldots$ recurrently if $\alpha'_{(0)}, \alpha''_{(0)}, \ldots$ defined by the bifurcation equation (8.57) are known. Each solution of this equation belongs to the space $\mathcal{A}^d$ and in accordance with the recurrence sequence of equations (8.62) it determines $\rho_{(0)} \in \mathcal{P}^r$.

Taking into consideration the one-to-one correspondence between the spaces $\mathcal{P}^r, \mathcal{A}^d$ and subspaces $P^r, A^d \subset {}^{n+1}$, we conclude that each solution of the bifurcation equation determines the component of the vector x in A^d, which in turn determines, by (8.62), the component of x in P^r. Following Thompson, the subspace A^d and space $\mathcal{A}^d$ will, therefore, be termed active, and P^r and $\mathcal{P}^r$ passive.

By differentiating Eqs. (8.49) with respect to λ, we obtain a recurrence system of equations analogous to system (8.59). We simplify it taking into account the results following from (8.51) and (8.61)

$$V^0_{,\rho} = 0, \qquad V^0_{,\alpha} = 0, \qquad \rho'_{(0)} = 0. \qquad (8.63)$$

As a result, we obtain a sequence of equations

$$V^0_{,\alpha\alpha}\alpha'_{(0)}\alpha'_{(0)} = 0,$$

$$3V^0_{,\rho\alpha}\rho''_{(0)}\alpha'_{(0)} + 3V^0_{,\alpha\alpha}\alpha''_{(0)}\alpha'_{(0)} + V^0_{,\alpha\alpha\alpha}\alpha'_{(0)}\alpha'_{(0)}\alpha'_{(0)} = 0. \qquad (8.64)$$

These equations, combined with system (8.62), enable one to determine successively the vectors $\alpha'_{(0)}, \rho''_{(0)}, \alpha''_{(0)}, \ldots$ and thus to determine, by expansion (8.58), the vector x in the neighborhood of a singular point. Equations (8.64) may have several solutions, and each of these solutions determines its own solution branch of the original system (8.1). In general, Eqs. (8.62), (8.64) contain the same information about the behavior of the solution in the neighborhood of a singular point as Eqs. (8.52), (8.53). However, they frequently permit the solution of the branching problem not on the basis of the bifurcation equation (8.57) but using its approximate (and simpler) representations. Some simple cases are discussed below.

3.   The simplest case of branching (rank()=n-1)

Let $rank(J^0) = n - 1$. In this case the dimensionality d of the active subspace A^d is 2. After transformations (8.43), (8.45), (8.51) with the use of the orthonormal bases of the subspaces P^{n-1} and A^2, the original system of equations (8.1) at a singular point reduces to

$$U_i(\rho_1, \ldots, \rho_{n-1}, \alpha_1, \alpha_2) = 0, \qquad i = \overline{1, n-1}, \qquad (8.65)$$

$$V(\rho_1, \ldots, \rho_{n-1}, \alpha_1, \alpha_2) = 0. \qquad (8.66)$$

Geometrically, this situation means that the active subspace A^2, which is a plane in $^{n+1}$, is in contact at any point with the solution set K (with all its branches passing through a singular point). The bifurcation analysis can, therefore, be reduced here to the bifurcation of a plane curve.

The first approximation of the bifurcation equation is the first of Eqs. (8.64). It becomes

$$V_{,11}^0(\alpha_1')^2 + 2V_{,12}^0\alpha_1'\alpha_2' + V_{,22}^0(\alpha_2')^2 = 0, \quad V_{,ij}^0 = \frac{\partial^2 V^0}{\partial\alpha_i\partial\alpha_j} \qquad (8.67)$$

which is a homogeneous quadratic form. Here, the following cases are possible:

1. *The quadratic form (8.67) is definite.* In this case it has a unique trivial solution $\alpha_1' = \alpha_2' = 0$. This means that in a close neighborhood of a singular point there are no more points from the desired solution set, i.e., the point under investigation is an isolated singular point. It is impossible to arrive at this point by moving along the continuous solution curve K. The appearance of an isolated singular point in the process of continuing the solution indicates, therefore, that the continuation process is not properly constructed. The reason of such a situation is usually an excessively large step for the continuation parameter λ. As previously mentioned in the chapter 1, the definiteness of the quadratic form (8.67) depends on the sign of its discriminant

$$D = V_{,11}^0 V_{,22}^0 - (V_{,12}^0)^2. \qquad (8.68)$$

If $D > 0$, form (8.67) is definite.

2. *The quadratic form (8.67) is indefinite.* In this case $D < 0$. If, for example, $V_{,22}^0 \neq 0$, the position of a tangent to the solution curve can be

defined on the plane $A^2 : (\alpha_1, \alpha_2) \subset {}^{n+1}$ by its angle φ with the α_1 axis. Then

$$t = tg\varphi = \alpha_2'/\alpha_1' = d\alpha_2/d\alpha_1. \tag{8.69}$$

From (8.67), we easily obtain a quadratic equation for t

$$V_{,11}^0 + 2V_{,12}^0 t + V_{,22}^0 t^2 = 0. \tag{8.70}$$

Since $D < 0$, this equation has two real roots

$$t_{1,2} = (-V_{,12}^0 \pm \sqrt{-D})/V_{,22}^0. \tag{8.71}$$

Thus, two solution branches intersect at a singular point whose tangents are defined by the expressions

$$d\alpha_2/d\alpha_1 = t_1, \qquad d\alpha_2/d\alpha_1 = t_2. \tag{8.72}$$

Geometrically this situation is shown in Fig. 8.1 in the plane A^2.

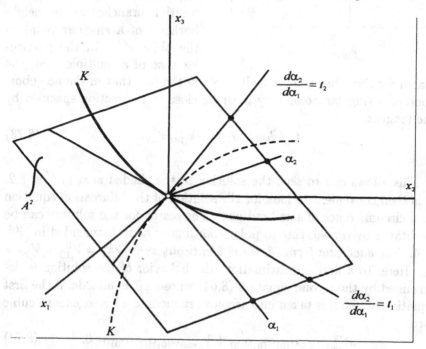

Figure 8.1.

3. *The quadratic form (8.67) is semi-definite.* In this case $D = 0$ and the quadratic trinomial (8.67) has a multiple root $t_{1,2} = t = -V_{,12}^0/V_{,22}^0$. To clarify the nature of a singular point requires analysis of higher order terms of the expansion and a more accurate analysis of the bifurcation equation. Examples of this kind of analysis for plane curves are given, for example, in [40], and analysis of possible cases is presented in [78, 114, 50]. Here, the singular point may turn out to be a point of contact of two solution branches or a cusp. In the latter case, the foregoing equations (8.62), (8.64) should be used with caution since they are developed on the assumption of the differentiability of the set of solutions with respect to λ at a singular point, and this condition is not fulfilled at a cusp.

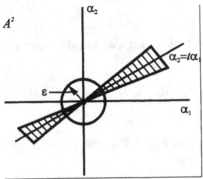

Figure 8.2.

For a numerical implementation of the continuation procedure at essentially singular point, the analysis of Eqs. (8.62), (8.64) for higher order derivatives does not appear to be convenient. A possible alternative is the numerical determination of the number and nature of solution branches in the neighborhood of a singular point in the plane A^2. In the particular case of a multiple root, the search for these branches is facilitated by the fact that in the neighborhood of a singular point they must be close to a direction specified by the tangent

$$t = d\alpha_2/d\alpha_1 = -V_{,12}^0/V_{,22}^0. \tag{8.73}$$

This allows one to seek the solution in the shaded area in Fig. 8.2. It is then convenient to look for the solution of the bifurcation equation on a circumference of small radius ε. The search for the solution can be facilitated by transferring to polar coordinates as recommended in [40].

4. *The quadratic form (8.67) is identically zero* ($V_{,11}^0 = V_{,12}^0 = V_{,22}^0 = 0$). Here, to a first approximation, the behavior of the solution is determined by the second equation (8.64) whose left-hand side, if the first equation of (8.62) is taken into account, reduces to a homogeneous cubic form

$$3V_{,\rho\alpha}^0 U_{,\alpha\alpha} \alpha_{(0)}' \alpha_{(0)}' \alpha_{(0)}' + V_{,\alpha\alpha\alpha}^0 \alpha_{(0)}' \alpha_{(0)}' \alpha_{(0)}' = 0. \tag{8.74}$$

By the argument used in the analysis of Eq. (8.67), we arrive at the conclusion that three solution branches may intersect at a singular point.

As has just been mentioned, the analysis of the bifurcation equation which takes into account higher order derivatives is not very suitable for numerical computation. It is more convenient to reduce the search for the solution branches to finding zeroes on the ε-circumference in the plane $A^2 : (\alpha_1, \alpha_2) \subset {}^{n+1}$ (Fig. 8.2).

4. The case of branching when rank()=n-2

To indicate the problems arising in the analysis of solution branches in more complicated cases, consider a situation when $r = rank(J^0) = n - 2$. Here the dimension d of active subspace A^d is 3. The bifurcation equations, to the second-order terms in the Taylor series expansion, are given by the first equation in (8.64). For $d = 3$, they assume the form in the following system of equations:

$$V_{1,jk}^0 \alpha_j' \alpha_k' = 0, \quad V_{2,jk}^0 \alpha_j' \alpha_k' = 0, \qquad j, k = 1, 2, 3. \tag{8.75}$$

The left-hand sides of these equations are homogeneous quadratic forms whose matrices are denoted as follows:

$$V^{(1)} = [V_{ij}^{(1)}] = [V_{1,jk}^0], \quad V^{(2)} = [V_{ij}^{(2)}] = [V_{2,jk}^0]. \tag{8.76}$$

Equations (8.75) in matrix form can be written as

$$(\alpha')^T V^{(1)} \alpha' = 0, \quad (\alpha')^T V^{(2)} \alpha' = 0, \quad \alpha' = \begin{bmatrix} \alpha_1' \\ \alpha_2' \\ \alpha_3' \end{bmatrix}. \tag{8.77}$$

It is clear that if at least one of these quadratic forms is definite, the singular point is isolated since in this case the system of equations (8.77) has no real solutions except for the trivial solution $\alpha' = 0$. But an isolated singular point cannot be reached if the continuation process is properly formulated. The definiteness of at least one of the quadratic forms (8.77) usually indicates that the step taken along the solution curve K is too large.

Consider some special cases when the quadratic forms (8.77) are not definite. In these cases, to each equation of (8.77) in the space $\mathcal{A}^3$ corresponds a real set of solutions. The problem of solving system (8.77) reduces to that of finding the intersection of these sets.

Let the eigenvalues of the matrices $V^{(i)}$, $i = 1, 2$, be denoted by $\lambda_1^{(i)}, \lambda_2^{(i)}, \lambda_3^{(i)}$, and the associated normalized eigenvectors by $s_1^{(i)}, s_2^{(i)}, s_3^{(i)}$. Consider various possible combinations of eigenvalues.

1. *The case* $\lambda_1^{(i)} > 0, \lambda_2^{(i)} > 0, \lambda_3^{(i)} < 0, \ i = 1, 2.$ Here the quadratic forms (8.77) are indefinite and their matrices each have two positive eigenvalues and one negative eigenvalue. Let us pass in the space $\mathcal{A}^3$: $\{\alpha_1, \alpha_2, \alpha_3\}$ to a basis formed by the eigenvectors $s_1^{(1)}, s_2^{(1)}, s_3^{(1)}$ of the matrix $V^{(1)}$. In other words, we use a transformation

$$\alpha' = p_1 s_1^{(1)} + p_2 s_2^{(1)} + p_3 s_3^{(1)} = S^{(1)} p,$$

$$p = (p_1, p_2, p_3)^T, \quad S^{(1)} = [s_1^{(1)}, s_2^{(1)}, s_3^{(1)}].$$

(8.78)

Here $S^{(1)}$ is a matrix whose columns are the eigenvectors $s_1^{(1)}, s_2^{(1)}, s_3^{(1)}$. By the orthonormality of the eigenvectors, the matrix $S^{(1)}$ is orthogonal, i.e., $S^{(1)T} S^{(1)} = E$, where E is the identity matrix. The first equation of (8.77) then becomes

$$(\sum_{i=1}^{3} p_i s_i^{(1)T}) V^{(1)} (\sum_{i=1}^{3} p_i s_i^{(1)}) = \sum_{i=1}^{3} p_i s_i^{(1)T} \sum_{i=1}^{3} p_i \lambda_i^{(1)} s_i^{(1)} =$$

$$= \lambda_1^{(1)} p_1^2 + \lambda_2^{(1)} p_2^2 - |\lambda_3^{(1)}| p_3^2 = 0.$$

(8.79)

Geometrically this means that in the space $\mathcal{A}^3$ the first of Eqs. (8.77) defines a surface in the form of an elliptic cone with vertex at the origin. The axis of the cone is directed along the eigenvector $s_3^{(1)}$. In exactly the same way, it can be shown that the second equation of (8.77) also defines an elliptic cone with axis along the vector $s_3^{(2)}$.

Thus, the equation of the real solutions of the system of equations (8.77) geometrically reduces to determining the common generators of two elliptic cones with common vertex at a singular point. To solve this, we apply transformation (8.78) to Eqs. (8.77). They take the form

$$\lambda_1^{(1)} p_1^2 + \lambda_2^{(1)} p_2^2 - |\lambda_3^{(1)}| p_3^2 = 0,$$

$$p^T P p = 0, \quad P = S^{(1)T} V^{(2)} S^{(1)}.$$

(8.80)

To simplify further the first of Eqs. (8.80), we set

$$\bar{p}_i = p_i / \sqrt{|\lambda_{(i)}^{(1)}|}, \quad i = 1, 2, 3,$$

(8.81)

which is equivalent to the following matrix operation:

$$\bar{p} = \overline{\Lambda}^{(1)} p,$$

(8.82)

$$p = \begin{bmatrix} \bar{p}_1 \\ \bar{p}_2 \\ \bar{p}_3 \end{bmatrix}, \quad \overline{\Lambda}^{(1)} = \begin{bmatrix} 1/\sqrt{\lambda_{(1)}^{(1)}} & 0 & 0 \\ 0 & 1/\sqrt{\lambda_{(2)}^{(1)}} & 0 \\ 0 & 0 & 1/\sqrt{|\lambda_{(3)}^{(1)}|} \end{bmatrix}.$$

The first of Eqs. (8.80) then assumes the simplest form

$$\bar{p}_1^2 + \bar{p}_2^2 - \bar{p}_3^2 = 0. \tag{8.83}$$

The structure of the second equation is basically unchanged, and becomes

$$\bar{p}^T P \bar{p} = 0, \qquad \overline{P} = \overline{\Lambda}^{(1)} P \overline{\Lambda}^{(1)}. \tag{8.84}$$

As a result of the above transformations, one of the cones under consideration becomes circular (8.83) and its axis in the space $\{\bar{p}_1, \bar{p}_2, \bar{p}_3\}$ coincides with the $\bar{p}_3$ axis.

The second cone is now defined by Eq. (8.84). In general, it remains elliptic, and is determined by the eigenvalues of the matrix $\overline{P}$. In particular, its axis is directed along the eigenvector which corresponds to the negative eigenvalue. The transformations performed are such that the number of positive and negative eigenvalues of the matrices $V^{(2)}$ and $\overline{P}$ is the same.

Now let the plane $\bar{p}_3 = 1$ be passed through the cones. The line of intersection of this plane with cone (8.83) is a circle, and for cone (8.84) it is an ellipse or a hyperbola depending on the relative positions of the cones. Thus, the problem of determining the real roots of Eqs. (8.77) has reduced to that of finding the common points of a unit circle and an ellipse or a hyperbola on the plane $\bar{p}_3 = 1$. The analytic solution of this problem in the general case does not apparently exist. Possible cases of relative positions of a circle and an ellipse are presented in Fig. 8.3. It is seen from this figure that the number of real solutions of Eqs. (8.77) may vary from 0 to 4. In particular, it is worth pointing out that the indefiniteness of the quadratic forms having the matrices $V^{(1)}$ and $V^{(2)}$ does not guarantee, by itself, the existence of real solutions of Eqs. (8.77). Such a situation occurs for the cases shown in Figs. 8.3(a), 8.3(b), and 8.3(c).

A final judgment of branching is only possible in the cases shown in Figs. 8.3(j) and 8.3(k). In the former, two solution branches intersect at the branch point. In so doing they touch two common generators of

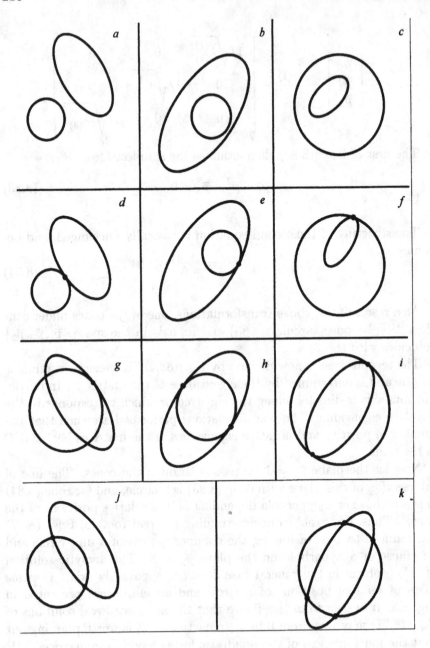

Figure 8.3.

cones (8.83), (8.84). In this case, therefore, continuation of the solution from the singular point is possible along four directions, as shown in

Fig. 8.4. In the plane passing through the common generators of the cones, the picture of branching is shown in Fig. 8.5.

In the case (k), four solution branches intersect at the singular point, touching four common generators of cones (8.83), (8.84). Here the branches no longer lie in one plane, and continuation of the solution from the singular point is possible in eight directions.

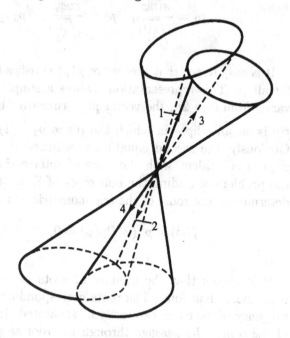

In the cases (d)-(i) (fig. 8.3), the cones have points of tangency. Here the corresponding common generator of the cones may turn out to be a tangent of two or more so-

Figure 8.4.

lution branches touching at the singular point. To find these solutions, it is necessary to consider the bifurcation equations, taking into account higher order terms in the Taylor expansion, in a plane tangent to both cones along the common generator.

The latter circumstance simplifies the analysis since the number of variables is reduced and becomes equal to two.In other words, in the space of active variables a two-dimensional subspace is singled out for each generator along which cones (8.83) and (8.84) are touching, namely a plane tangent to both cones along the common generator. It is in this subspace that the bifurcation equation is to be studied

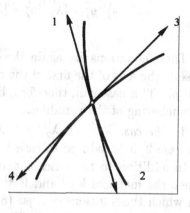

Figure 8.5.

If the solutions of Eqs. (8.80) are sought numerically, the following representation can be used for the vector $p = (p_1, p_2, p_3)^T$

$$p_1 = \frac{sin\varphi}{\sqrt{\lambda_1^{(1)}}}, \quad p_2 = \frac{cos\varphi}{\sqrt{\lambda_2^{(1)}}}, \quad p_3 = \frac{1}{\sqrt{|\lambda_3^{(1)}|}}. \tag{8.85}$$

It is easy to see that the vector $p(\varphi)$ satisfies the first equation (8.80) for all φ. This representation realizes a simple geometrical idea: as φ varies from 0 to 2π, the vector $p(\varphi)$ runs over the first cone so that its tip is on an ellipse in which the plane $p_3 = 1/\sqrt{|\lambda_3^{(1)}|}$ cuts this cone. Obviously, the second equation is satisfied if, and only if, the vector $p(\varphi)$ is coincident with the lines of intersection of the cones. Thus, the problem of finding the real roots of Eqs. (8.80) reduces to that of determining the roots of the trigonometric equation

$$f(\varphi) = p^T(\varphi)Pp(\varphi) = 0, \qquad 0 \le \varphi < 2\pi. \tag{8.86}$$

It is known that the number of roots in the interval $0 \le \varphi < 2\pi$ is no more than four. The roots corresponding to the intersection and tangency of the cones can easily be separated. In the case of intersection of the cones, the passage through the root as φ varies is accompanied by a sign change in the function $f(\varphi)$, while in the case of contact sign is unchanged.

2. *The case* $\lambda_1^{(1)} > 0, \lambda_2^{(1)} < 0, \lambda_3^{(1)} < 0, \lambda_1^{(2)} > 0, \lambda_2^{(2)} > 0, \lambda_3^{(2)} < 0$. By transformation (8.78), problem (8.77) is reduced to the following system of equations:

$$\lambda_1^{(1)}p_1^2 - |\lambda_2^{(1)}|p_2^2 - |\lambda_3^{(1)}|p_3^2 = 0, \quad p^TPp = 0. \tag{8.87}$$

These equations are again the equations of cones but, in contrast to case 1, the axis of the first of the cones is directed along the axis p_1, and not p_3. This case can, therefore, be reduced to case 1 by an appropriate renumbering of the variables.

3. *The case* $\lambda_1^{(1)} > 0, \lambda_2^{(1)} < 0, \lambda_3^{(1)} < 0, \lambda_1^{(2)} > 0, \lambda_2^{(2)} < 0, \lambda_3^{(2)} < 0$. As for case 2, it is reduced to case 1 by renumbering the variables.

In addition to the cases considered above, there may be situations when the matrices $V^{(1)}$ and $V^{(2)}$ have zero eigenvalues. The sets in $\mathcal{A}^3$ on which the solutions of Eqs. (8.77) lie degenerate either into straight lines or into planes. This facilitates the search for the real solutions of the system of equations (8.77). A great number of various combinations are possible. As an example, consider one of these.

4. *The case* $\lambda_1^{(1)} > 0, \lambda_2^{(1)} = 0, \lambda_3^{(1)} < 0, \lambda_1^{(2)} > 0, \lambda_2^{(2)} > 0, \lambda_3^{(2)} < 0$. Transformation (8.78) reduces system (8.77) to the following:

$$\lambda_1^{(1)} p_1^2 - |\lambda_3^{(1)}| p_3^2 = 0, \quad p^T P p = 0. \tag{8.88}$$

The solution set of the second equation in the space $\{p_1, p_2, p_3\}$ forms a cone as before. The solution of the first equation form two planes

$$p_1 = \pm \sqrt{\frac{|\lambda_3^{(1)}|}{\lambda_1^{(1)}}} p_3. \tag{8.89}$$

Substitution of each of these relations in the second equation of (8.88) reduces the problem to a homogeneous quadratic equation in p_2 and p_3 of the form

$$C_1 p_2^2 + 2C_2 p_2 p_3 + C_3 p_3^2 = 0. \tag{8.90}$$

Thus, in each of the planes (8.89) in $\mathcal{A}^3$ the problem of finding solution branches reduces to that considered in the previous section.

In conclusion it may be noted that, when the dimension of the space of active variables $d > 2$, it is necessary to consider a number of various situations to find solution branches at a singular point. This makes the analysis of solutions at a singular point cumbersome and very time-consuming in practice. The general theory of this problem is far from completion. For more detailed information we refer to the monographs [117, 3, 62]. In practice, it is often efficient to use perturbed solutions. An example of such an approach is the numerical analysis on the behavior of a three-bar system presented in Section 1.4.

References

[1] Ahlberg J.H., Nilson E.N. and Walsh J.L. The theory of splines and their applications. Academic Press, New York, London. 1967.

[2] Alexandrov A.Ya. and Solov'ev Yu.I. Three - dimensional problems of elasticity theory [in Russian]. Nauka, Moscow. 1978.

[3] Arnol'd V.I., Varchenko A.N. and Gusein-Zade S.M. Singularities of differentiable maps. Birkhauser, Basel. 1985.

[4] Arushanyan O.B. and Zaletkin S.F. Numerical methods for ordinary differential equations implemented in FORTRAN. [in Russian]. Moscow, Mosk. Gos. Univ. 1990.

[5] Baker C. T. H. Methods for integro-differential equations // Numerical solution of integral equations. Oxford: Clarendon Press, 1974. P. 189-206.

[6] Baker C. T. H. The numerical treatment of integral equations. Oxford: Clarendon Press, 1976.

[7] Bakhvalov N.S. Numerical methods. [in Russian]. V. 1. Nauka, Moscow. 1973.

[8] Bellman R.E. and Kalaba R.E. Quasilinearization and nonlinear boundary - value problems. Elsevier, N.-Y., 1965.

[9] Belotserkovskii S.M., Nisht M.I., Ponomarev A.T. and Rusev O.V. Computer investigation of parachutes and hang-gliders. [in Russian]. Moscow, Mashinostroenie, 1987.

[10] Bergan P.G., Horrigmoe G., Krakeland B., Söreide T.H. Solution techniques for nonlinear finite element problems // Int. J. Num. Meth. Eng. 1978. V. 12. 12. P. 1677 – 1696.

[11] Boyarinstev Yu.E., Danilov V.A., Loginov A.A. and Chistyakov V.F. Numerical methods for solving singular systems. [in Russian]. Novosibirsy, Nauka. 1989.

[12] Brauer A. Limits for the characteristic roots of matrix // Duke Math. J. 1946. N 13. P. 387-395.

[13] Brenan K. E., Campbell S. L., Petzold L. R. Numerical solution of initial-value problems in differential-algebraic equations. N. -Y., Amsterdam, L.: North-Holland, 1989.

[14] Bruder J., Strehmel K., Weiner R. Partitioned adaptive Runge-Kutta methods for the solution of nonstiff and stiff systems // Numer. Math. 1988. V. 52. P. 621-638.

[15] Byrne G. D., Hindmarsh A. C. A polyalgorithm for the numerical solution of ordinary differential equations // ACM. Trans on Math. Software. 1975. V. 1. P. 71-96.

[16] Byrne G. D., Hindmarsh A. C. Stiff ODE solvers: A review of current and coming attraction // J. of Computational Physics. 1987. V. 70. N 1. P. 1-62.

[17] Campbell S. L. Singular system of differential equations. San- Francisco, L., Melbourn: Pitman Advanced Publ. Program., 1980.

[18] Campbell S. L. Singular system of differential equations. II. San- Francisco-L. -Melbourn: Pitman Advanced Publ. Program., 1982.

[19] Courant R., Fridrichs K., Lewy H. Ueber die partiellen Differenzen-gleichungen der mathematischen Physik // Math. Ann. 1928. V. 100. P. 32-74.

[20] Cryer C. W. Numerical methods for functional differential equations // Delay and functional differential equations and their applications. Proc. of the Park City Conf. N. -Y.: Acad. Press, 1972. P. 17-102.

[21] Curtis A. R. The FACSIMILE numerical integrator for stiff initial value problems // AERE-R. 9352. Oxfordshire:AERE Harwell, 1978.

[22] Curtiss C. F., Hirschfelder J. O. Integration of stiff equations // Proc. of the National Academy of Sciences of US. 1952. V. 38. P. 235-243.

[23] Dahlquist G. A special stability problem for linear multistep methods // BIT. 1963. N 3. P. 27-43.

[24] Davidenko D.F. On a new method for the numerical solution of systems of nonlinear equations // Dokladi Academii Nauk of Russia, V. 88. 4. P. 601-602.

[25] Davidenko D.F. On the approximate solution of systems nonlinear equations // Ukr. Mat. Zh. 1953. V. 5. 2. P. 196-206.

[26] Davidenko D.F. The application of the method of variation of the parameters to the construction of iterative formulas of increased accuracy for the numerical solution of nonlinear integral equations // Dokladi Academii Nauk of Russia, V. 162. 3. P. 499 – 502.

[27] Dorodnitsin A.A. asymptotic solution of Van-der-Pole equation // Prikl. Mat. Mekh. V.11. 3. P. 313-322. 1947.

[28] Doolan E.P., Miller J.J.H. and Schilders W.H.A. Uniform numerical methods for problems with initial and boundary layers. Boole Press. Dublin. 1980.

[29] El'sgol'ts L.E. On the approximate integration of differential equations with retarded argument // Prikl. Mat. Mekh. V.15. 4. P. 771-772. 1951.

[30] El'sgol'ts L.E. The approximate methods of integration of differential - difference equations // Uspekhi Mat. Nauk. 1953. V. 8. 4. P. 81 - 93.

[31] El'sgol'ts L.E. and Norkin S.B. An introduction to the theory of differential equations with retarded argument. [in Russian]. Nauka, Moscow. 1971.

[32] Faux L.D. and Pratt M.J. Computational geometry for design and manufacture. Ellis Horwood, Chichester. 1979.

[33] Fedorenko R.P. To regular stiff systems of ordinary differential equations // Dokladi Academii Nauk of Russia. 1983. V. 273. 6. P. 1318-1322.

[34] Fedorenko R. P. Stiff systems of ordinary differential equations // Numerical processes and systems. Vol. 8. 1991. P. 328-380.

[35] Fedorenko R. P. Stiff systems of ordinary differential equations // Numerical methods and applications. Boca Raton, Ann Arbor, L., Tokyo: CRC Press. 1994. P. 117-154.

[36] Feldstein A. Discretization methods for retarded ordinary differential equations // Ph. D. Thesis. Los Angeles: Univ. of California, 1964.

[37] Feldstein A., Sopka J. R. Numerical methods for nonlinear Volterra integro differential equations // SIAM J. Numer. Anal. 1974. V. 11. P. 826-846.

[38] Ficken F. The continuation method for nonlinear functional equations // Comm. Pure Appl. Math. 1951. V. 4. 4. P. 435 - 456.

[39] Field J. R., Noyes R. M. Oscillations in chemical systems. IV, Limit cycle behavior in a model of a real chemical reaction // J. Chem. Physics. 1974. V. 60. P. 1877-1884.

[40] Fikhtengol'ts G.M. A course of differential and integral calculus. Vol. 1. [in Russian]. Nauka, Moscow. 1969.

[41] Gavurin M.K. Nonlinear functional equations and continuous analogs of iterative methods // Izv.Vyssh. Uchebn. Zaved., Mat., N 5. P.18-31. 1958.

[42] Gear C. W. Numerical initial value problems in ordinary differential equations. N. - Y. :Prentice Hall, Englewood Cliffs, 1971.

[43] Gear C. W. The simultaneous numerical solution of differential- algebraic equations // IEEE Trans. Circuit Theory. CT.-18. 1971. P. 89-95.

[44] Gerasimov B.P. and Kul'chitskaya I.A. STIFSP package of programs for integrating differential - algebraic systems of large dimension. Preprint Inst. Appl. Math., Usssr Acad. Sci., Moscow. 1984. 103.

[45] Gershgorin S.A. Uber die Abqrenzung der Eigenwerte einer Matrix. Izv. Academii Nauk SSSR, Ser. fiz.-mat., P. 749-754. 1931.

[46] Godunov S.K. Numerical solution of boundary value problems for systems of linear ordinary differential equations // Uspekhi Mat. Nauk. 1962. V. 16. 6. P. 171 - 174.

[47] Griepentrog E., Marz R. Differential-algebraic equations and their numerical treatment. Leipzig: Teubner, 1986.

[48] Grigolyuk E.I., Shalashilin V.I. Problems of nonlinear deformation. Kluwer Academic Publishers, Dordrecht / Boston / London, 1991.

[49] Grigorenko Ya. M. and Mukoed A.P. Computerized solution of nonlinear problems in the theory of shells. [in Russian]. Vishcha shkola, Kiev. 1983.

[50] Gulyaev V.I., Bazhenov V.A. and Gotsulyak E.A. Stability of nonlinear mechanical systems. [in Russian]. Vishcha shkola, L'vov. 1982.

[51] Hairer E., Norsett S.P., Wanner G. Solving ordinary differential equations. I. Springer—Verlag, Berlin, 1987.

[52] Hairer E., Lubich C., Roche M. The numerical solution of differential-algebraic systems by Runge-Kutta methods. Berlin etc.: Springer, 1989.

[53] Hairer E., Wanner G. Solving ordinary differential equations 2. Stiff and differential-algebraic problems. Berlin, e. a. :Springer- Verlag, 1991.

[54] Hamming R.W. Numerical methods for scientists and engineers. Graw-Hill Book Com. Y.-N. 1962.

[55] Henrici P. Discrete variable methods in ordinary differential equations. N. -Y. :Wiley, 1962.

[56] Hindmarsh A. C. LSODE and LSODI, two new initial value ordinary differential equations solvers // ACM. SIGNUM. Newsletter. 1980. V. 15. N 4. P. 10-11.

[57] Hindmarsh A. C. ODEPACK // A systematized collection of ODE solvers in numerical methods for scientific computation. N. - Y. : North- Holland. 1983. P. 55-64.

[58] Joss G. and Joseph D. Elementary stability and bifurcation theory. Springer - Verlag, New York etc. 1980.

[59] Karmishin A.V., Lyaskovets V.A., Myachenkov V.I. and Frolov A.N. Statics and dynamics of thin-walled shell structures. [in Russian]. Mashinostroenie, Moscow. 1975.

[60] Kisner W. A numerical method for finding solutions of nonlinear equations // SIAM J. Appl. Math. 1964. V. 12. P. 424 – 428.

[61] Kleinmichel H. Stetige Analoge und Iterations verfaher für nichtlinear Gleichungen in Banachräumen // Math. Nach. 1968. V. 37. P. 313 – 314.

[62] Krasnosel'skii M.A., Vainikko G.M., Zabreiko P.P., Rutitskii Ya.B. and Stestenko V.Ya. Approximate solution of operator equations. [in Russian]. Nauka, Moscow. 1969.

[63] Kulikov G.Yu. Numerical solution of an autonomous Cauchy problem with an algebraic constraint on phase variables // Zh. Vychisl. Mat. Mat. Fiz. V. 33. 4. P. 522-540. 1993.

[64] Kuznetsov E.B. and Shalashilin V.I. Cauchy's problem as a problem of the continuation of a solution with respect to a parameter // Zh. Vychisl. Mat. Mat. Fiz. V. 33. 12. P. 1792-1805. 1993.

[65] Kuznetsov E.B. and Shalashilin V.I. The Cauchy problem for deformable systems as a parametric solution continuation problem // Izv RAN. Mekh Tverd Tela. V. 28. 6. P. 145-152. 1993.

[66] Kuznetsov E.B. and Shalashilin V.I. Cauchy's problem as a problem of continuation with respect to the best parameter // Diff. Urav. V.30. 6. P. 964-971. 1994.

[67] Kuznetsov E.B. and Shalashilin V.I. The Cauchy problem for mechanical systems with a finite number of degrees of freedom as a problem of continuation on the best parameter // Prikl. Mat. Mekh. V.58. 6. P. 14-21. 1994.

[68] Kuznetsov E.B. and Shalashilin V.I. Pellet charge motion along a gun barrel // Izv RAN. Mekh Tverd Tela. V. 29. 1. P. 189-199. 1994

[69] Kuznetsov E.B. and Shalashilin V.I. A parametric approximation // Zh. Vychisl. Mat. Mat. Fiz. V. 34. 12. P. 1757-1769. 1994.

[70] Lahaye M. E. Une metode de resolution d'une categorie d'equations transcendentes // Compter Rendus hebdomataires des seances de L'Academie des sciences. 1934. V. 198. 21. P. 1840-1842.

[71] Lahaye M. E. Solution of system of transcendental equations // Acad. Roy. Belg. Bull. Cl. Sci. 1948. V. 5. P. 805-822.

[72] Lambert J. D. Computational methods in ordinary differential equations. N. -Y. : Wiley, 1973.

[73] Lebedev A.A. and Chernobrovkin L.S. Dynamics of flight. [in Russian]. Moscow, Mashinostroenie, 1973.

[74] Lebedev V. I. How to solve stiff systems of differential equations by explicit methods // Numerical processes and systems. Vol. 8. 1991. P. 237-291.

[75] Lebedev V. I. How to solve stiff systems of differential equations by explicit methods // Numerical methods and applications. Boca Raton, Ann Arbor, L., Tokyo:CRC Press. 1994. P. 45-80.

[76] Lur'c A.I. Analytical Mechanics. [in Russian]. Moscow, Fizmatgiz. 1961.

[77] Lyapunov A.M. General problem of motion stability. [in Russian]. Kharkov, 1892.

[78] Lyttleton R.A. The stability of rotating liquid masses. Cambrige: University Press. 1953.

[79] Marchuk G.I. and Lebedev V.I. Numerical methods in neutrons transfer theory. [in Russian]. Moscow, Atomizdat, 1981.

[80] Marcus M., Minc H. A survey of matrix theory and matrix inequalities. Allynd and Bacon, Inc, Boston, 1964.

[81] Modern numerical methods for ordinary differential equations. Edited by G. Hall and J.M. Watt. Oxford. Clarendon Press. 1976.

[82] Morozov N.F. Nonlinear theory of thin plates // Dokladi Academii Nauk of Russia. 1957. V. 114. 5. P. 968 – 971.

[83] Morozov N.F. Nonlinear problems in theory of thin plates // Vestn. Leningr. Univ. 1958. 19. P. 100 – 124.

[84] Morozov N.F. Uniqueness of the symmetric solution of a large deflection problem for a symmetrically loaded circular plate // Dokladi Academii Nauk of Russia. 1958. V. 123. 3. P. 417 – 419.

[85] Morozov N.F. On the existence of a nonsymmetric solution in a large deflection problem for a circular plate under a symmetric load // Izv. Vyssh. Uchebn. Zaved., Mat. 1961. 2. P. 126 – 129.

[86] Morozov N.F. Nonlinear problems in the theory of thin plates with p axes of symmetry // Tr. Leningr. Tekhol. Inst. Tsellyul.-Bumazh. Prom., 1962. 11. P. 206 – 208.

[87] Nazarenko N.A. Approximation of plane curves by parametric Hermitian splines // Ukr. Mat. Zh. V. 31. 2. P. 201-205. 1979.

[88] Oran E.S. and Boris J.P. Numerical simulation of reactive flow. Elsevier, New York etc. 1987.

[89] Ortega J.M. and Poole W.G. An introduction to numerical methods for differential equations. Pitman Publishing inc., 1981.

[90] Pavlov N.N. and Skorospelov V.A. Modelling curves and surfaces in a system for automating geometric calculations // Spline Functions in Engineering Geometry. Computational Systems. 86. P. 44-59. Inst. Mat. SO Akad. Nauk SSSR. Novosibirsk. 1981.

[91] Petrov V.V. Finite deflection analysis of shallow shells // Nauch. Dokl. Vyssh. Shkoly., Stroit. 1959. 1. P. 27-35.

[92] Petrovskii I.G. Lectures on theory of ordinary differential equations. [in Russian]. Nauka, Moscow. 1970.

[93] Petzold L. R. A description of DASSL: A differential-algebraic system solver // Scientific Computing. Amsterdam: North-Holland, 1983. P. 65.

[94] Poincare H. Sur l'equilibre d'une masse fluide animal d'un mouvement de rotation// Acta mathem. 1885. V. 7. P. 259 - 380.

[95] Rakitskii Yu.V., Ustinov S.M. and Chernorutskii I.G. Numerical methods for solving stiff systems. [in Russian]. Nauka, Moscow. 1979.

[96] Rentrop P. Partitioned Runge-Kutta methods with stiffness detection and stepsize control // Numer. Math. 1985. V. 47. P. 545-564.

[97] Ridel V.V. and Gulin B.V. Dynamics of soft shells. [in Russian]. Nauka, Moscow. 1990.

[98] Riks E. The application of Newton's method to the problem of elastic stability // Trans. ASME. J. Appl. Mech. 1972. V. E39. 4. P. 1060-1065.

[99] Riks E. A unified method for the computation of critical equilibrium states of nonlinear elastic systems // Acta Techn. Acad. Sci. Hung. 1978. V. 87. 1-2. P. 121-141.

[100] Riks E. An incremental approach to the solution of snapping and buckling problems // Int. J. Solids Struct. 1979. V. 15. 7. P. 529-551.

[101] Riks E. Some computational aspect of the stability analysis of nonlinear structures // Comput. Math. Appl. Mech. Engrg. 1984. V. 47. 3. P. 219-259.

[102] Samarskii A.A. and Gulin A.V. Numerical methods. [in Russian]. Nauka, Moscow. 1989.

[103] Schmidt E. Zur Theorie linearen und nichtlinearen Integralgleichungen. Teil 3. Über die Auflösungen der nicht-linear Integralgluchungen und die Verveigung ihrer Lösunger// Math.Ann. 1908. S. 370 - 399.

[104] Sendov B. Hausdorff approximations. Bulgarian Academy of Sciences, Sofia. 1979.

[105] Shalashilin V.I. The continuation method and its application to a large deflection problem for a nonshallow circular arch // Izv. Akad. Nauk SSSR. Mekh. Tverd. Tela. 1979. 4. P. 178 - 184.

[106] Shalashilin V.I. and Kuznetsov E.B. Cauchy problem for nonlinear deformation of systems as a parametric continuation problem // Dokladi Academii Nauk of Russia, v.329, N 4, P. 426–428. 1993.

[107] Shalashilin V.I. and Kuznetsov E.B. To formulation of Cauchy problem for the systems with concentrated parameters // Problems of machine - building and reliability of machines. 1994. 3. P. 120-121.

[108] Shalashilin V.I. and Kuznetsov E.B. The best parameter in the continuation of a solution // Dokladi Academii Nauk of Russia, V.334, N 5, P. 566–568. 1994.

[109] Shampine L. F., Gear C. W. A user's view of solving stiff ordinary differential equations // SIAM Review. 1979. V. 21. N 1. P. 1-17.

[110] Sherman A. H., Hindmarsh A. C. GEAR: A package for the solution of sparse, stiff ordinary differential equations // Electrical Power Problems. The mathematical challenger. SIAM. Philadelphia. 1980. P. 190.

[111] Shimanskii V.E. Computerized numerical solution of boundary value problems. [in Russian]. Part. 2. Naukova Dumka, Kiev. 1966.

[112] Sincovec R. F., Erisman A. M., Yip E. L., Epton M. A. Analysis of descriptor system using numerical algorithms // IEEE Trans. on Auto Control. 1980. V. 26. P. 139-147.

[113] Thompson J.M.T., Hunt G.W. A general theory of elastic stability. London: G. Wiley interscience publ. 1973.

[114] Thompson J.M.T., Hunt G.W. Towards a unified bifurcation theory // J. Appl. Math. and Physis. 1975. V. 26. P. 581 - 603.

[115] Tikhonov A.N. and Samarskii A.A. Equations of mathematical physics. [in Russian]. Nauka, Moscow. 1972.

[116] Trenogin V.A. Lusternic's theorem and the best parametrization of solutions of nonlinear equations // Functional analysis and applications. 1998. V. 38. 1 P. 87-90.

[117] Vainberg M.M. and Trenogin V.A. Theory branching of the solutions of non-linear equations. [in Russian]. Nauka, Moscow. 1969.

[118] Vakarchuk S.B. Approximation of curves given in parametric form by means of spline curves // Ukr. Mat. Zh. V. 36. 2. P. 352-355. 1983.

[119] Vasil'eva A.B. and Butuzov V.F. Asymptotic expansions of solutions of singu-lar - perturbed equations [in Russian]. Nauka, Moscow. 1973.

[120] Voevodin V.V. and Arushanyan O.B. Numerical analysis in FORTRAN. [in Russian]. Moscow, Mosk. Gos. Univ. 1979.

[121] Vorovich I.I. and Zipalova V.F. Solution of nonlinear boundary value elasticity problems by passing to a Cauchy problem // Prikl. Mat. Mekh. 1965. V.29. 5. P. 894-901.

[122] Weiner R., Strehmel K. A type insensitive code for delay differential equations basing on adaptive and explicit Runge- Kutta interpolation methods // Com-puting. 1988. V. 40. P. 255-265.

[123] Widlund O. B. A note on unconditionally stable linear multistep methods // BIT. 1967. V. 7. P. 65-70.

[124] Willard L. Numerical methods for stiff equations. N. -Y. :Acad. Press, 1981.

[125] Zav'yalov Yu.S., Kvasov B.I. and Miroshnichenko V.L. Spline - functions meth-ods. [in Russian]. Nauka, Moscow. 1980.